DATE DUE

Demco, Inc. 38-293

Volume 77

Gene-Environment Interplay

Advances in Genetics, Volume 77

Serial Editors

Theodore Friedmann
University of California at San Diego, School of Medicine, USA

Jay C. Dunlap
The Geisel School of Medicine at Dartmouth, Hanover, NH, USA

Stephen F. Goodwin
University of Oxford, Oxford, UK

Volume 77

Gene- Environment Interplay

Edited by

Marla B. Sokolowski

Department of Ecology and Evolutionary Biology and
The Fraser Mustard Institute for Human Development
University of Toronto
Toronto, Ontario, Canada

Stephen F. Goodwin

Department of Physiology, Anatomy and Genetics
University of Oxford
Oxford, United Kingdom

AMSTERDAM • BOSTON • HEIDELBERG • LONDON
NEW YORK • OXFORD • PARIS • SAN DIEGO
SAN FRANCISCO • SINGAPORE • SYDNEY • TOKYO
Academic Press is an imprint of Elsevier

ELSEVIER

Academic Press is an imprint of Elsevier

525 B Street, Suite 1900, San Diego, CA 92101-4495, USA
225 Wyman Street, Waltham, MA 02451, USA
32 Jamestown Road, London, NW1 7BY, UK
Radarweg 29, POBox 211, 1000 AE Amsterdam, The Netherlands

First edition 2012

ISBN: 978-0-12-387687-4
ISSN: 0065-2660

For information on all Academic Press publications
visit our website at store.elsevier.com

Printed and bound in USA

12   13   14   11   10   9   8   7   6   5   4   3   2   1

Contents

4   The Circadian Clock of the Fly: A Neurogenetics Journey Through Time   79

Özge Özkaya and Ezio Rosato

Contributors

Numbers in parentheses indicate the pages on which the authors' contributions begin.

Jade Atallah (59) Department of Biology, University of Toronto at Mississauga, Mississauga, Ontario, Canada

Guy Bloch (1) Department of Ecology, Evolution and Behavior, The Alexander Silberman Institute of Life Sciences, The Hebrew University of Jerusalem, Jerusalem, Israel

Frances A. Champagne (33) Department of Psychology, Columbia University, New York, USA

Ada Eban-Rothschild[1] (1) Department of Ecology, Evolution and Behavior, The Alexander Silberman Institute of Life Sciences, The Hebrew University of Jerusalem, Jerusalem, Israel

Joel D. Levine (59) Department of Biology, University of Toronto at Mississauga, Mississauga, Ontario, Canada

Özge Özkaya (79) Department of Genetics, University of Leicester, Leicester, United Kingdom

Ezio Rosato (79) Department of Genetics, University of Leicester, Leicester, United Kingdom

Jonathan Schneider (59) Department of Biology, University of Toronto at Mississauga, Mississauga, Ontario, Canada

[1]Current address: Department of Psychiatry and Behavioral Sciences, Stanford University, Stanford, California, USA

Preface

There are a number of common themes that have emerged in recent years from both animal and human research on genes and behavior. The concept of gene–environment interplay is one of the most exciting. Our thinking about how individual differences in behavior arise has come a long way from the days when psychologists answered this question by saying it is the environment and biologists attributed it to genes. Fortunately, this nature–nurture dichotomy has finally been put to rest. Even the question of what proportion of phenotypic variation is genetic compared to environmental (genes + environment) makes little sense with our current understanding about how genes are able to listen to the environment. The concept of gene–environment interactions (G × E) emerged from the inadequacies of broad sense heritability measures. Initially G × E were determined by statistical interactions between behavioral measures of two or more genotypes measured in two or more environments. With the advent of genetic technologies, research moved onto measuring G × E effects arising from allelic variation in single genes in individuals exposed to different environments. More recently, G × E have been measured at the genome level.

The concept of gene–environment interplay encompasses both G × E and epigenetics. Epigenetics offers another set of biological mechanisms whereby the environment can get under our skin without the necessity of there being DNA sequence variation. It provides biological mechanisms for why identical twins with the same DNA sequences nevertheless differ in their phenotypes. Like G × E, changes in the epigenome are responsive to many interesting environmental factors within and around the individual including the social environment. We now know that our genome is listening to our environment and responds to it with changes in the expression of our genes.

Gene–environment interplay changes during development. There is an active interplay during prenatal and postnatal development and adulthood between our genome and the environments we experience. What happens at one development stage significantly affects the developmental events that follow. Research has come a long way from the now debunked concept of nature–nurture.

This volume provides a sampling of some of the new and elegant research being performed on gene–environment interplay.

Chapter 1 is from Ada Eban-Rothschild and Guy Bloch; they review studies in *Drosophila melanogaster* and the bee *Apis mellifera*, which suggest that social influences on the clock are more important than previously recognized. Studying social behavior in *Drosophila* has all the advantages of advanced molecular genetics with anatomical, physiological, and behavioral approaches, and the bee can be studied and manipulated in an ecologically relevant context. These studies will be a stepping stone to important breakthroughs in our understanding of the mechanisms underlying social influences on sleep and circadian rhythms.

Chapter 2 is from Frances A. Champagne. The relatively new field of epigenetics, the study of how a gene's function or expression can be changed without alterations to the genetic code, has changed the way we think about our genetic makeup. In Champagne's fascinating review, she discusses the epigenetics of social experience, emphasizing the role of epigenetic mechanisms in shaping variation in social behavior and the implications of epigenetics for our understanding of the transmission of traits across generations.

Chapter 3 is from Joel D. Levine's lab. Levine's group has pioneered studies in *Drosophila* by observing fly social interactions. They have noted that in group situations flies engage in complex social interactions that influence learning, circadian clocks, aggression, and mating. This systems-level network approach to the study of *Drosophila* groups will be key to understanding the regulation and modulation of such group social behaviors and the importance of context in regulating them.

Chapter 4 is from Ezio Rosato's lab. This fresh perspective on the circadian clock in the fly provides a uniquely deep understanding of the topic under discussion amenable to all readers, while at the same time bringing in those papers that challenge the current notion of how the molecular clock operates at the cellular level, and provides a new view on the organization and the evolution of circadian clocks.

MARLA B. SOKOLOWSKI
STEPHEN F. GOODWIN

1

Social Influences on Circadian Rhythms and Sleep in Insects

Ada Eban-Rothschild[1] and Guy Bloch

Department of Ecology, Evolution and Behavior, The Alexander Silberman Institute of Life Sciences, The Hebrew University of Jerusalem, Jerusalem, Israel

I. Introduction
II. The Circadian System of Insects
III. Social Influences on Circadian Rhythms in Insects
    A. Social synchronization
    B. Social influences on the ontogeny of circadian rhythms
    C. Social modulation of plasticity in the expression of circadian rhythms
    D. Additional social influences on circadian rhythms
IV. Sleep
    A. Sleep in invertebrates
    B. Sleep in honey bees with and without circadian rhythms
V. Social Influences on Sleep in Insects
VI. Conclusions
    Acknowledgments
    References

[1]Current address: Department of Psychiatry and Behavioral Sciences, Stanford University, Stanford, California, USA

Advances in Genetics, Vol. 77
0065-2660/12 $35.00
http://dx.doi.org/10.1016/B978-0-12-387687-4.00001-5

ABSTRACT

The diverse social lifestyle and the small and accessible nervous system of insects make them valuable for research on the adaptive value and the organization principles of circadian rhythms and sleep. We focus on two complementary model insects, the fruit fly *Drosophila melanogaster*, which is amenable to extensive transgenic manipulations, and the honey bee *Apis mellifera*, which has rich and well-studied social behaviors. Social entrainment of activity rhythms (social synchronization) has been studied in many animals. Social time givers appear to be specifically important in dark cavity-dwelling social animals, but here there are no other clear relationships between the degree of sociality and the effectiveness of social entrainment. The olfactory system is important for social entrainment in insects. Little is known, however, about the molecular and neuronal pathways linking olfactory neurons to the central clock. In the honey bee, the expression, phase, and development of circadian rhythms are socially regulated, apparently by different signals. Peripheral clocks regulating pheromone synthesis and the olfactory system have been implicated in social influences on circadian rhythms in the fruit fly. An enriched social environment increases the total amount of sleep in both fruit flies and honey bees. In fruit flies, these changes have been linked to molecular and neuronal processes involved in learning, memory, and synaptic plasticity. The studies on insects suggest that social influences on the clock are richer than previously appreciated and have led to important breakthroughs in our understanding of the mechanisms underlying social influences on sleep and circadian rhythms.   © 2012, Elsevier Inc.

I. INTRODUCTION

The social environment, which we define as the sum of social interactions, cues, and signals encountered by an animal, has a profound influence on its development, physiology, and behavior. In the current review, we focus on the influences of the social environment on two fundamental biological systems, circadian rhythms and sleep, in insects. Circadian rhythms are biological rhythms that cycle with a daily period of roughly 24 h (see Section II). Sleep is a complex behavioral state that is characterized by reduced activity and responsiveness (see Section IV). These two systems interact in various ways. For example, the circadian clock influences the timing of sleep and wakefulness, and clock genes are implicated in the regulation of different aspects of sleep, such as sleep quality and homeostasis (Andretic et al., 2005; Sehgal and Mignot, 2011). Circadian rhythms and sleep are sensitive to environmental variables such as light and temperature (for a recent review see Peschel and Helfrich-Forster, 2011). There is also evidence that circadian rhythms and sleep are influenced by the social environment, but little is known about the functional significance or the mech-anisms underlying these interactions (Davidson and Menaker, 2003; Mistlberger and Skene, 2004). Insects provide an excellent system for addressing these questions because their nervous system is relatively small and accessible. Insects are very diverse in terms of their activity phase and include species that are diurnal, nocturnal, or crepuscular, as well as those that can switch between these states or be active with no circadian rhythms. Importantly, insects show a broad range of social lifestyles, ranging from solitary to highly structured societies.

Highly eusocial insects such as ants, honey bees, and termites are characterized by the restriction of reproduction to a single or a few individuals (e.g., queens), by the fact that individuals from older generations cooperatively care for the young, and by an elaborate communication system (Wilson, 1971). The social environment influences almost every facet of the life of highly social insects, including their patterns of activity and sleep (see below). The social environment is also important for many animals living in simple societies, and for solitary animals as well. The social environment of these animals may include interactions with potential mates, offspring, or conspecifics with which they compete for resources such as territory, shelter, or mates. We focus mainly on the highly social honey bee and the fruit fly, which is commonly considered solitary or facultatively gregarious. The honey bee provides an excellent model for identifying and studying the social signals influencing circadian rhythms and sleep in an ecological context. The arsenal of genetic and transgenic toolkits available for *Drosophila* make it an excellent model for studying the molecular mechanisms underlying the interaction between social factors and the circadian and sleep systems.

II. THE CIRCADIAN SYSTEM OF INSECTS

Circadian rhythms are defined as biological rhythms that meet the following three criteria: (1) they persist, or "*free-run*," with a period of about 24 h in the absence of external time cues, (2) they are reset, or *entrained*, by environmental cues, in particular, light and temperature, and (3) they exhibit "*temperature compensation*"; in other words, their period length is stable over a wide range of physiological temperatures. The circadian clock influences many physiological and behavioral processes in insects. These include activity, the sleep-wake cycle, feeding, mating, oviposition, egg hatching, and pupal eclosion. The circadian clock is also involved in measuring day length, and influences photoperiodism and annual rhythms, such as diapause and seasonal reproduction (Dunlap *et al.*, 2004; Saunders, 2002).

The circadian system has been traditionally described as having three functional components, although today this description is regarded as a useful conceptual framework rather than a strict description of the system

(Zhang and Kay, 2010). The *core* of the clock is composed of *pacemakers*, cell-autonomous rhythm generators that cycle approximately with a 24-h period. The central pacemaker is entrained by *input pathways* in which environmental signals are detected, converted to sensory information, and transmitted to the central pacemakers. *Output pathways* carry temporal signals away from pacemaker cells to various biochemical, physiological, and behavioral processes (Bell-Pedersen *et al.*, 2005; Dunlap *et al.*, 2004; Saunders, 2002). The molecular bases of rhythm generation in pacemakers of organisms as diverse as fungi, plants, fruit flies, and mammals consist of interlocked autoregulatory transcriptional/ translational feedback loops with positive and negative elements. The pacemaker cells of animals are interconnected in a circadian network that couples their activities and orchestrates normal rhythms in physiology and behavior (Bell-Pedersen *et al.*, 2005; Dunlap *et al.*, 2004).

The insect for which the circadian clockwork has been best characterized is the fruit fly *Drosophila melanogaster* (see Peschel and Helfrich-Forster (2011) for a recent review, and Zhang and Kay (2010) for an updated list of clock genes). The transcription factors *Clock* (*Clk*) and *Cycle* (*Cyc*) (the *positive elements*) form a dimer and activate the transcription of the transcription factors *Period* (*Per*) and *Timeless* (*Tim1*) (the *negative elements*). *Per* and *Tim1* are translated into proteins that enter the nucleus where they interfere with the transcriptional activity of the CLK:CYC complex and concomitantly shut down their own expression. *Par Domain Protein 1* (*Pdp1*), *Vrille* (*Vri*), and *Clockwork Orange* (*Cwo*) act together with *Clk* in an interlocked feedback loop that is thought to stabilize the *Per/Tim1* loop. Additional proteins including several kinases and phosphatases fine tune this cell-autonomous rhythm generation machinery by modulating the stability of the canonical clock genes. Drosophila-type *Cryptochrome* (*Cry-d*, also known as *insect Cry1*) has a photic input function. Although the genes and the organizational principles of the molecular clockwork of mammals and *Drosophila* are similar, there are some important differences between them. For example, mammals do not have orthologs for *Tim1* and *Cry-d*, but have three paralogs for *Per*. They also have two paralogs for *Cry* (mammalian-type *Cry*, their insect orthologs are also known as insect *Cry2*) that act together with the *Per* genes in the negative loop of the clock.

The recent sequencing of the genomes of several insects and other invertebrates indicates that there is notable variation in the molecular organization of the insect clock. For example, the genome of the honey bee and other species of the order Hymenoptera does not encode orthologs to *Cry-d* and *Tim1* genes, but does encode the mammalian-type transcriptional repressing *Cry-m* (Ingram and BloIngram *et al.* submitted; Rubin *et al.*, 2006; Yuan *et al.*, 2007). The functional significance of this variability in the molecular organization of the circadian clockwork in insects is unknown (Rubin *et al.*, 2006).

There is also considerable variation in the anatomical organization of the circadian network in the insect brain (Helfrich-Forster *et al.*, 1998; Sehadova *et al.*, 2003; Zavodska *et al.*, 2003). The clock network of *Drosophila* comprises about 150 neurons/brain hemisphere that are typically classified into seven major neuroanatomical groups. There are three groups located in the dorsal part of the brain, termed dorsal neurons 1–3 (DN_{1-3}). The other four groups are located more laterally and ventrally and are called lateral neurons (LN_d, l-LN_v, LPN, and s-LN_v). Clock proteins are also expressed in a few hundred glia cells, but little is known about their importance for chronobiological functions (Helfrich-Foerster *et al.*, 2007; Nitabach and Taghert, 2008). Immunostaining with antibodies directed against clock proteins suggests that there is a wide range of species-specific variation in the anatomical organization of the circadian network in insects (e.g., Helfrich-Forster *et al.*, 1998; Sehadova *et al.*, 2003; Zavodska *et al.*, 2003). For example, in the honey bee, both the PER and *Pigment Dispersing Factor* (PDF, a peptide in the circadian system of insects)-immunoreactive clusters are located in brain areas that have been implicated in the regulation of circadian rhythms in *Drosophila*, but currently there is no evidence for cells expressing both PER and PDF as in *Drosophila* (Bloch *et al.*, 2003; Sehadova *et al.*, 2004).

III. SOCIAL INFLUENCES ON CIRCADIAN RHYTHMS IN INSECTS

The social environment of insects consists of conspecific individuals at all life stages (i.e., eggs, larvae, pupae, and adults), which they may contact directly or indirectly by sensing cues and signals they release to the environment (such as volatile pheromones or acoustic signals). Such social factors in the environment may influence the phase (entrainment), strength, expression, or development of circadian rhythms.

A. Social synchronization

Social synchronization, or "social entrainment" is the best studied social influence on the circadian clock. Social synchronization in animals experiencing various social interactions (including courtship, aggression, and caging with conspecifics) has been tested for diverse species and experimental protocols (Table 1.1). These studies, mainly with mammals, suggest that animals differ considerably in their entrainment by social time givers. Some species are entrained well even by cues such as volatile odorants and sounds that do not require direct contact (e.g., Goel and Lee, 1997; Levine *et al.*, 2002; Rajaratnam and Redman, 1999; Southwick and Moritz, 1987), while other species fail to show social entrainment even after forceful social interactions such as overt aggression or sexual experience (e.g., Gattermann and Weinandy, 1997;

Table 1.1. Studies Testing Social Entrainment in Animals

Animal species	Order, class	Degree of sociality	Social interactions	Entrainment	Reference
Fruit fly (*Drosophila melanogaster*)	Diptera, insecta	Solitary/facultative gregarious	Contact with other males Sexual contact with a female	Fair No	Levine *et al.* (2002), Krupp *et al.* (2008) Fujii *et al.* (2007)
Madeira cockroach (*Leucophaea maderae*)	Dictyoptera, insecta	Gregarious	Contact with conspecifics	No	Knadler and Page (2009)
Western honey bee (*Apis mellifera*)	Hymenoptera, insecta	Highly eusocial	Contact with other workers	Good	Frisch and Koeniger (1994), Moritz and Kryger (1994), Southwick and Moritz (1987)
			Contact with the queens	Good	Moritz and Sakofski (1991)
Golden hamster (*Mesocricetus auratus*)	Rodentia, mammalia	Solitary	Acoustical and olfactorial communication	No	Gattermann and Weinandy (1997)
			Pulses of social interactions	No	Refinetti *et al.* (1992)
			Contact with other males	No	Refinetti *et al.* (1992)
			Pulses of social interactions	Variable	Mrosovsky (1988)
Indian Palm Squirrel (*Funambulus palmarum*)	Rodentia, mammalia	Pair to group living	Contact with other males	Fair	Rajaratnam and Redman (1999)
Mongolian gerbil (*Meriones unguiculatus*)	Rodentia, mammalia	Social	Acoustical and olfactorial communication	No	Gattermann and Weinandy (1997)
Common degu (*Octodon degus*)	Rodentia, mammalia	Social	Contact with conspecifics	Good for females, no for males	Goel and Lee (1995)
Sugar glider (*Petaurus breviceps*)	Marsupialia, mammalia	Social	Contact with the opposite sex	No	Kleinknecht (1985)
Rhesus monkey (*Macaca mulatta*)	Primates, mammalia	Social	Contact with conspecifics	Good	Yellin and Haury (1971)
Common marmoset (*Callithrix jacchus*)	Primates, mammalia	Social	Acoustic communication	Weak	Erkert *et al.* (1986)
Leschenault's rousettte (*Rousettus leschenaulti*)	Chiroptera, mammalia	Gregarious to social	Acoustic communication	Good	Vanlalnghaka *et al.* (2005)
Schneider's Leaf-nosed Bat (*Hipposideros speoris*)	Chiroptera, mammalia	Gregarious to social	Indirect contact	Good	Marimuthu *et al.* (1981)

The upper part summarizes studies on insects, and the lower part shows representative studies on mammals demonstrating the lack of a clear relationship

Refinetti *et al.*, 1992; see Table 1.1). This apparent variability contrasts with studies on the better explored photic signals that produce stable entrainment in the vast majority of species studied so far.

The relationships between social lifestyle and social entrainment are not straightforward. As can be expected, the highly social honey bee and the social Rhesus monkey, but not the solitary Golden hamster, show effective social entrainment (Table 1.1). On the other hand, social interactions fail to entrain circadian rhythms in the social Mongolian gerbil or Sugar glider. Similarly, the gregarious cockroach *Leucophaea maderae*, which exhibits a broad range of social interactions, failed to show social synchronization in experiments in which individuals were placed in various social settings (Knadler and Page, 2009). Social entrainment seems to be important in dark cavity-dwelling social animals such as the Western honey bee (*Apis mellifera*) and some species of bats. Social information is perhaps specifically important in these species because of the absence of potent time givers in their environment, such as light and temperature fluctuations (Table 1.1). Given that cavity nesting seems to be a derived trait in honey bees (Winston, 1987), it would be interesting to determine whether cavity dwelling and open nesting honey bee species differ in their entrainment by social *Zeitgebers*.

The results of studies concerning the efficacy of social entrainment of locomotor activity rhythms in *Drosophila* flies are contradictory. Male flies changed their temporal locomotor activity profile when monitored in the presence of a receptive female (Fujii *et al.*, 2007). However, when these males were transferred to individual cages they reverted to their normal activity pattern and phase, suggesting that in flies, sexual interactions do not produce stable entrainment of circadian rhythms in locomotor activity. On the other hand, there is evidence suggesting stable social entrainment in groups of male flies. Male wild-type flies that were placed in small groups for 19 days and then transferred to individual cages showed better phase coherence than flies that were isolated for the entire period in individual cages (Levine *et al.*, 2002). The social composition of the group had a profound effect: the phase was less coherent when *per0* mutants, which have no circadian rhythms, were present. Caging flies with individuals coming from a different "time zone" (that have a different phase of activity) produced significant phase shifts under constant conditions (Levine *et al.*, 2002), indicating that the phase of activity is affected by interactions between individual flies in the group. The variation in studies on *Drosophila* show that the efficacy of social entrainment depends strongly on the social context and experimental design. For example, some of this variability may be explained by differences in the size of groups (number of interacting flies) investigated in the different studies. A similar variation was also reported for the Golden hamster, which was socially entrained in some studies but not in others (Table 1.1). These multiple studies on *Drosophila* and hamsters highlight the need to perform a comprehensive set of experiments before concluding that social *Zietgebers* cannot entrain the clock of a certain species.

Studies on *Drosophila* have begun defining the molecular and neuronal mechanisms governing social entrainment. Video recordings of pairs of male or female flies revealed only a few events of close-proximity encounters, suggesting that direct contact is not the main form of interaction involved in social entrainment in this species (Fujii *et al.*, 2007). Consistent with this notion, individually isolated flies could be entrained by exposure to a flow of air passing through a vial with 10–15 flies, but not by air passing through a similar vial without flies (Levine *et al.*, 2002). Additional studies of mutant and transgenic flies have indicated that social entrainment is mediated by volatile pheromones and detected by the olfactory system. Pheromone detection in the olfactory system is regulated by the circadian clock and is involved in the social entrainment of locomotor activity rhythms (Krishnan *et al.*, 1999; Krupp *et al.*, 2008; Levine *et al.*, 2002). It should be noted that the gustatory response is also regulated by peripheral circadian clocks (Chatterjee *et al.*, 2010). It would be worthwhile testing whether contact pheromones that are licked or detected by other sensory organs are also involved in social synchronization. Social entrainment in male flies also appears to be mediated by circadian rhythms in pheromone production in the oenocytes, which are abdominal cells specialized in the production of cuticular hydrocarbons (Krupp *et al.*, 2008). This may explain why the presence of *per0* mutants that do not produce oscillating pheromonal signals interfered with the synchronization in a group of flies (Levine *et al.*, 2002).

Courtship and mating in *Drosophila* are also circadianally regulated and influenced by clock genes (Beaver and Giebultowicz, 2004; Hardelan, 1972; Sakai and Ishida, 2001; Tauber *et al.*, 2003). In the presence of a female, *Drosophila* males exhibit long periods of courtship activity with a pronounced rest phase at dusk. These rhythms continue in a constant environment indicative of their endogenous nature. The courtship intensity is controlled mainly by the male clock (Fujii *et al.*, 2007), but the female clock may have prime effects on the time of mating (female receptivity, Sakai and Ishida, 2001; Tauber *et al.*, 2003). The courtship rhythms thus contrast with the locomotor activity of isolated males that typically exhibit an activity peak at dusk (Fujii and Amrein, 2010). Experiments in which the olfactory system of flies was genetically or mechanically manipulated suggest that olfactory signals also play a key role in the social regulation of courtship rhythms (Fujii *et al.*, 2007). Genetic manipulations further indicate that courtship rhythms are mediated by PDF-expressing ventral lateral neurons and the DN_1 subset of dorsal neurons (Fujii and Amrein, 2010). The lateral neurons are sufficient for the male fly to show circadian rhythms in courtship behavior, while the DN_1 neurons are essential for phase synchronization. Fujii and Amrein (2010) further suggested that DN_1

neurons influence circadian rhythms in courtship behavior and locomotor activity by two distinct signaling pathways. Many clock neurons in the s-LN$_v$, LN$_d$, and DN$_1$ clusters express the male splice form of the *Fruitless* gene (FRU^M), and are therefore sexually dimorphic in physiology (Fujii and Amrein, 2010; Lee *et al.*, 2006). Given the evidence that FRU^M expressing neurons define the courtship circuitry in *Drosophila*, the expression of FRU^M in clock neurons may provide a neuroanatomical means of accounting for the observed influence of the clock on male courtship behavior.

Social entrainment can be expected to be important in insects living in complex societies. Many of these species nest in dark cavities and social synchronization is likely to be an efficient mechanism for coordinating their activities. An emerging property of social synchronization is a colony level circadian rhythm. Colony rhythms, like those of individuals honey bees, have a stable free running period in a constant environment and their phase can be shifted by changes in environmental factors such as light, temperature, and feeding (Frisch and Aschoff, 1987; Frisch and Koeniger, 1994; Kefuss and Nye, 1970; Moore, 2001). Ten-day-old honey bee workers that are removed from the colony and thus deprived of their time givers show a very coherent phase on the first few days in isolation but later drift from each other and from the colony phase (Frisch and Koeniger, 1994). Preforager honey bees ($\sim$2–3 weeks of age) that stay inside the environmentally regulated hive show circadian rhythms in their resting time, with a phase similar to that of foragers (Moore *et al.*, 1998). Workers at this age typically labor as "nectar receivers" that accept nectar loads collected by returning foragers, and therefore need to be synchronized with the timing of foraging (Crailsheim *et al.*, 1996). It is possible that preforager honey bees are entrained by environmental factors such as light and temperature during orientation flights (Capaldi *et al.*, 2000; Lindauer, 1961), or by temperature fluctuations at the hive periphery (which differs from the brood area that is tightly thermoregulated) (Winston, 1987). However, honey bees are in synchrony with the external environment even if restricted to the inner part of the hive and therefore deprived of any light or flight experience ("BigBack" bees; Bloch *et al.*, 2004). In particular, their brain *Per* mRNA levels are higher at night, as is typical of foragers that experience the external environment. Furthermore, when "BigBack" bees are removed from the hive and monitored individually, they are seen to have circadian rhythms in locomotor activity with higher activity during the subjective day. Interestingly, honey bees as young as 3 days of age, which are active around the clock in the colony, show circadian rhythms in locomotor activity and a coherent phase when removed from the hive (Eban-Rothschild and Bloch, in preparation; see Section III.B). A similar social synchronization was not shown by same-age sister bees that spent the same period in individual cages outside the hive (Eban-Rothschild and Bloch, in preparation). This indicates that honey bee workers do not emerge from the pupae with a synchronized circadian rhythm, unlike many insects that time their eclosion (Saunders, 2002), but rather are entrained by the colony environment.It should be noted that all the evidence for social entrainment are from studies with workers. It is not clear whether social synchronization is similarly effective for drones and gynes. The timing of mating flights in honey bees is species specific (Koeniger and Koeniger, 2000), and there is evidence suggesting that the syn-chronization of drones and gynes arriving from different colonies is mediated by environmental factors such as the light–dark cycle (Sasaki, 1990).

What in the colony environment mediates the phase synchronization between individual honey bee workers? There is evidence suggesting that both nestmate workers and the queen entrain circadian rhythms in workers (Moritz and Kryger, 1994; Southwick and Moritz, 1987). Young honey bees that were caged outside the hive for 2 days with 30 same-age bees showed weak or no phase coherence and their activity onset often did not occur during the early subjective day (Eban-Rothschild and Bloch, in preparation). This observation suggests that social contact with young group mates outside the context of the colony is not sufficient for effective synchronization. By contrast, honey bees that were caged in single- or double-mesh enclosures inside the hive showed strong phase coher-ence, similar to that of their sisters moving freely in the hive (Fig. 1.1; Eban-Rothschild and Bloch, in preparation). These observations suggest that strong synchronization of honey bees in the colony can be achieved without direct contact with other workers, the queen, or the brood. These findings are consis-tent with an earlier study in which rhythms in temperature and oxygen con-sumption were partially synchronized between two groups of workers separated by a solid Plexiglas partition (Moritz and Kryger, 1994). Synchronization was improved in experiments in which the Plexiglas division was punched with holes. Together these studies suggest that either fluctuations in the physical environment of the hive (e.g., humidity, temperature, or CO_2 concentration) or volatile pheromones are the prominent factors involved in the synchronization of honey bee workers.

B. Social influences on the ontogeny of circadian rhythms

Most insects emerge from the pupa with strong circadian rhythms. In many species, the time of eclosion is tightly gated by the clock (Saunders, 2002), and even some larval stages show clear circadian rhythms (e.g., Kostal et al., 2009; Sehgal et al., 1992). By contrast, studies on bumble bees (Yerushalmi et al., 2006), honey bees (Moore, 2001), and ants (Jong and Lee, 2008) suggest that in these insects, circadian rhythms develop only after the adult ecloses from the pupa. This development is reminiscent of the postembryonic ontogeny of circadian rhythms in infants of humans and other primates (Rivkees, 2003).

Place and duration	Treatment	Development of circadian rhythms	Social synchronization	Sleep amount
In a colony for 24 h	Freely moving	+	**ND**	+++
In a colony for 48 h	Freely moving	+++	+++	+++
	Enclosed in a single mesh-enclosure*	+++	+++	+++
	Enclosed in a double mesh-enclosure*	+++	+++	+++
In the lab for 48 h	Within a group of 30	+	+	++
	Individually isolated	+	−	+

Figure 1.1. The influence of different social environments on the development and synchronization of circadian rhythms and the amount of sleep in young honey bees. Newly emerged honey bees experienced the various environments for 24–48 h and were then transferred to individual cages in the lab in which their locomotor activity was monitored in constant conditions. *, similar results were obtained in two separate experiments in which bees were caged either individually or in a group of 30. The number of "+" signs is proportional to the amount of the measured variable; "−," no effect; "ND," not determined. Circadian rhythms and sleep were differentially affected by the colony environment. See text for details.

The ontogeny of circadian rhythms has been best explored in the honey bee in which young workers care for the brood and perform additional in-hive activities around the clock, with attenuated or no behavioral and molecular rhythms. Later in life, these bees show robust circadian rhythms that are associated with the transition to other activities such as nest guarding or foraging (see Section III.C below). Honey bees that are isolated individually or placed in small groups in the lab shortly after emergence also show an ontogeny of circadian rhythms; they do not show circadian rhythms in locomotor activity, metabolism, or clock gene expression in their first days as adults, but later show robust circadian rhythms (e.g., Meshi and Bloch, 2007; Spangler, 1972; Stussi and Harmelin, 1966; Bloch and Meshi 2007). The expression of circadian rhythms in the colony is modulated by the social environment. Nurse-bees, which are active and care for the brood around the clock, show strong circadian rhythms in locomotor activity and clock gene expression when they are removed from the hive or separated from the brood at 7 days of age (Shemesh et al., 2007, 2010). Even honey bees as young as 3 days of age that are active around the clock in the colony showed circadian rhythms in locomotor activity shortly after transfer to constant conditions in the lab (Eban-Rothschild et al., 2012). By contrast, sister bees isolated individually from emergence were not likely to show circadian rhythms at this age. These findings are somewhat counter-intuitive because in the colony young honey bees typically do not express circadian rhythms, but their circadian system still appears to develop faster in the colony than in isolation (Fig. 1.2).

What accounts for the more rapid development of the circadian system in the colony environment than in individual isolation? To answer this question, Eban-Rothschild et al. (2012) manipulated the environment of worker honey bees during the first 24–48 h postpupal eclosion. They found that bees experiencing the colony environment for only 24 h, or staying for 48 h with 30 same-age sister bees in a cage outside the hive, had weak circadian rhythms that were similar to those of individually isolated bees (Fig. 1.1). By contrast, honey bees that were caged individually, or with thirty other newly emerged bees in single- or double-mesh (preventing direct contact) enclosures inside the hive showed circadian rhythms comparable to bees of a similar age moving freely in the hive. The bees caged in the hive experienced the physical environment of the hive

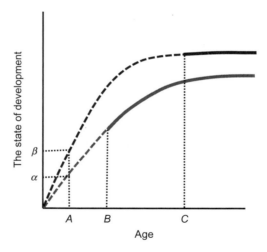

Figure 1.2. A schematic model for the development of the circadian system in honey bees in isolation or in a colony. The dashed line depicts the underlying development of the circadian system and the solid part of the curve depicts the expression of overt circadian rhythms in locomotor activity. Red—a model for bees isolated in the lab immediately after emergence; blue—a model for bees developing in a typical field colony. The development of the circadian system is faster in bees in the colony, but they naturally express overt circadian rhythms relatively late in life (age, $C > B$). When young honey bees are removed from the colony at an early age (A), they are more likely to show circadian rhythms than honey bees of a similar age that were removed from the colony immediately after emergence ($\alpha < \beta$). (See Color Insert.)

and were exposed to volatile odors, but could not move freely in the hive. These results indicate that direct contact with the queen, other workers, or the brood cannot account for the influence of the colony environment, because these were all impeded by the double-mesh enclosures. These results are striking because they suggest that important social signals such as contact pheromones and trophallaxis are not necessary for the colony to influence the ontogeny of circadian rhythms. Although light might have some influence on the ontogeny of circadian rhythms in honey bees (Bloch et al., 2001), it cannot account for these findings because the bees in all treatment groups were similarly exposed to daylight. Furthermore, light is not likely to regulate the development of the circadian system in field colonies in which young honey bees typically stay in the dark cavity of the nest. Hence, the ontogeny of circadian rhythms in young honey bees may be modulated by the social environment (i.e., volatile pheromones), the hive microenvironment (e.g., temperature, humidity, and CO_2 levels), or both. Additional studies are needed to evaluate the relative importance of each of these factors.

C. Social modulation of plasticity in the expression of circadian rhythms

Workers in several species of social bees and ants are reported to switch between activities with and without circadian rhythms as a function of the task they perform. Foragers typically have strong circadian rhythms, whereas nurse workers typically care for the brood around the clock with no circadian rhythms (reviewed in Bloch, 2009, 2010). This task-related chronobiological plasticity has been best studied in honey bees (A. mellifera) in which the division of labor relates to worker age (e.g., nurses are typically younger than foragers). This remarkable plasticity in circadian rhythms is thought to be functionally significant because it improves task performance and overall colony efficiency. Honey bee larvae are frequently tended by nurse workers (Huang and Otis, 1991; Shemesh et al., 2007, 2010), and therefore around-the-clock activity may enable nurses to better provide the brood needs. On the other hand, foragers collect nectar and pollen during the day and rely on the circadian clock to time their visits to flowers and for time-compensated sun-compass navigation. The hypothesis that plasticity in circadian rhythms is functionally significant is supported by the strong link between division of labor and circadian rhythmicity (e.g., foragers have strong rhythms and nurses are active around the clock in diverse colony demographics and regardless of age) (Bloch and Robinson, 2001; Moore et al., 1998), and by the evidence for a similar task-related plasticity in additional social insects. These include the bumble bee Bombus terrestris, in which division of labor is based primarily on size (Yerushalmi et al., 2006), and several species of ants in which division of labor evolved independently from that of the honey bee (Ingram et al., 2009; Jong and Lee, 2008).

Given that brood care is the main activity of nurse honey bees, and that the brood appears to benefit from continuous care, the most straightforward hypothesis is that worker activity is regulated by signals from the brood. This hypothesis is supported by the findings that nurse-age bees that were caged in broodless combs inside or outside the hive showed strong circadian rhythms in activity and clock gene expression (Shemesh et al., 2010). Nurse honey bees without the flagella of their antenna cared for the brood with strong circadian rhythms whereas their full-sisters, sham-treated bees (whose antennae were touched but not cut), cared for the brood around-the-clock similar to nurses in typical field colonies (Nagari and Bloch, in press). Flagella-less and sham-treated bees showed similar circadian rhythms in locomotor activity when analyzed individually in constant lab conditions, indicating that the operation by itself did not affect the circadian clock. These observations suggest that the flagella are involved in mediating the brood signal, which modulates plasticity in circadian rhythms in honey bees (Nagari and Bloch, in press). The identity of the brood signal(s) and the possible involvement of additional sensory modalities are yet to be determined.

Plasticity in circadian rhythms was also found in bumble bee (B. terrestris) queens in which it appears to be associated with maternal behavior or physiology (Eban-Rothschild et al., 2011). The bumble bee queen is active around the clock at the colony founding stage, in which she cares alone for the developing brood. However, queens whose brood was removed typically switched to activity with strong circadian rhythms. Interestingly, some of the queens switched back to around-the-clock activity before actually laying again, suggesting that maternal behavior or physiology rather than the brood influenced plasticity in circadian rhythms (Eban-Rothschild et al., 2011). The ovaries and the endocrine signals they secrete (e.g., ecdysteroids) do not seem to regulate this plasticity because a similar effect on circadian rhythmicity was also observed in ovaryectomized queens (Eban-Rothschild et al., 2011). The influence of the ovaries on circadian rhythms in worker honey bees and bumble bees is yet to be determined.

Another important line of research addresses the mechanisms underlying plasticity in the circadian system. Several hypotheses may explain this remarkable plasticity. First, the differences in activity profiles may be associated with the fact that foragers experience strong variation in light and temperature whereas the environment of nurses is overall constant. However, this hypothesis is not consistent with the evidence that nurses are also active around the clock under a light–dark illumination regime, and that foragers continue to show strong circadian rhythms in constant conditions (Moore, 2001; Rubin et al., 2006; Shemesh et al., 2007, 2010). The second line of hypotheses relates to task-related variation in age or development. For example, the circadian system of nurses that are typically young could be undeveloped or underdeveloped.

However, nurses switch to activity with strong circadian rhythms shortly after transfer to the laboratory, indicating that their circadian system is capable of generating robust rhythms (Shemesh *et al.*, 2007, 2010). In addition, old foragers with strong circadian rhythms may revert to care for the brood around the clock, like nurses in normal colonies (Bloch and Robinson, 2001; Bloch *et al.*, 2001). Third, pacemakers in the nurse brain may generate normal molecular oscillations but do not affect behavioral rhythms because they are masked or uncoupled from motor controlling centers. This hypothesis that predicts similar oscillations in brain clock gene expression in nurses and foragers is not consistent with the evidence that clock gene oscillations are attenuated in nurses (Bloch *et al.*, 2001, 2004; Shemesh *et al.*, 2007; Toma *et al.*, 2000). Fourth, the molecular feedback loop in pacemaker cells in the nurse brain could be fixed at a certain state. This hypothesis predicts that the molecular and behavioral cycling should resume from the point when the bee is released from the hive environment. In other words, if the nurse is removed from the hive, the phase of oscillations outside the hive would be correlated with the time of removal. When this hypothesis was explicitly tested it was not supported by the data; the onset of activity was correlated with the subjective morning in the hive from which the nurses were collected, and not with the time of removal from the hive (Bloch, 2010; see Fig. 1.3.2 in Eban-Rothschild and Bloch, 2012). The fifth hypothesis that seems to be the one best supported by the available data states that the function or organization of the circadian network differs between nurses active around the clock and rhythmic foragers. For example, it is possible that some pacemakers in the brain of nurses active around the clock generate circadian rhythms but with a different phase. It is also possible that only a few pacemakers oscillate in the nurse brain, whereas cycling stops in most other clock cells. The pacemaker cells are synchronized with each other again when the nurse is removed from the hive or switches to activities with little or no direct contact with the brood. Finally, oscillators in the nurse brain may cycle, but with a low amplitude that is not sufficient to drive circadian rhythms in locomotor activity (Shemesh *et al.*, 2010). A better description of the neuroanatomy of the honey bee circadian network is needed to test these alternative hypotheses.

Genome-wide expression analyses are consistent with the hypothesis that some processes are circadianally regulated in nurses active around the clock. Rodriguez-Zas *et al.* (2012) used a high density oligonucleotide probe microarray to compare gene expression in nurses and foragers collected around the clock from self-sustained colonies that were each placed in an observation hive. They found 160 and 541 transcripts that exhibited significant sinusoidal oscillations in nurses and foragers, respectively. These oscillating transcripts had their peak expression distributed throughout the day in both foragers and nurses (Fig. 1.3A). The microarray results for transcripts involved in circadian rhythms and visual systems were consistent with earlier qPCR and Northern blot analyses

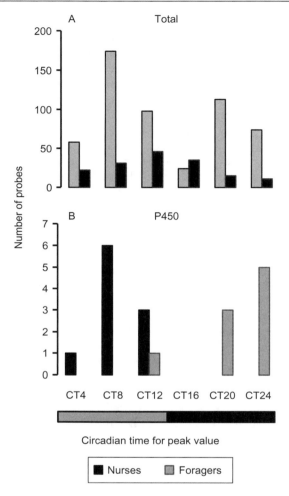

Figure 1.3. Number of probes showing oscillations in transcript abundance during the day in nurse and forager honey bee workers. (A) Total number of oscillating transcripts. Overall, 541 and 160 transcripts oscillated in foragers and nurses, respectively. (B) Only probes encoding putative Cytochrom P450 genes. Nurses and foragers were entrained for 3–6 days in a 12-h light:12-h dark illumination regime and collected in constant darkness. The horizontal bar at the bottom shows the illumination regime: Black, subjective night; gray, subjective day (data modified from Rodriguez-Zas *et al.*, 2012).

for the same genes (Sasagawa *et al.*, 2003; Shemesh *et al.*, 2007, 2010). Gene ontology analyses suggested enrichment for oscillating genes from the functional categories "development" and "response to stimuli" (foragers), "muscle contraction" and "microfilament motor gene expression" (nurses), and "genera-tion of precursor metabolites" and "energy" (both). Transcripts of genes encoding P450 enzymes oscillated in both nurses and foragers but with a different phase (Fig. 1.3B). Additional studies are needed to confirm transcript oscillation in nurses and to identify the cells in which they are expressed.

D. Additional social influences on circadian rhythms

Krupp et al. (2008) reported that in groups of male flies exposed to arrhythmic per0 mutants, not only was the phase coherence weaker (see Section III.A above), but the overall levels or the amplitude of clock gene oscillations in the oenocyte or head were also reduced. Thus, the social environment of the fly appears to affect both the central and peripheral clocks. These changes in clock gene expression were associated with alterations in the overall amount or the temporal profile of cuticular hydrocarbon secretion, and therefore could affect the odor of male flies. Interestingly, these changes were associated with differ-ences in mating success: wild-type flies in heterogeneous groups in which two of the six males were per0 mutant mated more with their six female partners than males in homogenous groups with six male and six female wild-type flies. It is difficult to interpret these findings in the framework of the natural behavioral ecology of fruit flies (e.g., it could stem from the reduced mating success of the mutant flies). Nevertheless, these results suggest that social influences on the circadian system of Drosophila are not limited to synchronization of the phase of activity. In a broader sense, these studies with flies suggest that the influences of the social environment on the circadian system are diverse in many insects, and are not limited to highly social species such as honey bees and ants.

IV. SLEEP

Sleep is a fundamental and evolutionarily conserved biological phenomenon. Sleep research has traditionally focused on humans and other mammals, and it was commonly believed that invertebrates do not sleep in the strict sense. This view has been revised dramatically over the past two decades during which sleep-like states have been described for diverse nonmammalian species, including fish (Prober et al., 2006), insects (Hendricks et al., 2000; Kaiser and Steiner-Kaiser, 1983; Tobler, 1983), and even nematode worms (Raizen et al., 2008). The molecular pathways associated with sleep in mammals, flies, fish, and worms show much conservation, suggesting an ancient and common origin for sleep (Allada and Siegel, 2008; Cirelli, 2009). The three main areas of molecular conservation in the pathways regulating sleep in vertebrates and invertebrates are the involvement of circadian clock genes (such as *Per*), signaling pathways (such as EGF receptor) and genes involved in neurotransmission (such as GABA receptors) (Allada and Siegel, 2008; Cirelli, 2009).

Sleep is commonly defined by three main behavioral criteria: (1) a period of quiescence associated with a species-specific posture and/or resting place, which is typically accompanied by reduced motor activity; (2) an elevated response threshold (i.e., a stronger stimulus is needed to produce a response); and (3) a homeostatic regulation mechanism, which is manifested in a sleep rebound after periods of sleep deprivation (Hendricks *et al.*, 2000; Tobler, 1983). Sleep is distinguished from quiet wakefulness by the reduction in the ability to react to stimuli, while the reversibility to an awake state distinguishes sleep from coma (Siegel, 2005).

Sleep is also associated with a set of pharmacological, electrophysiological, and molecular characteristics. The electroencephalogram (EEG) pattern (which corresponds to neuronal activity in the brain) is commonly used to define sleep behavior in studies of mammals and birds (in combination with the recordings of muscular activity). Changes in EEG patterns during sleep correspond with two sleep states: "rapid eye movement" (REM) sleep and "slow wave sleep" (SWS or non-REM). In humans, non-REM (NREM) sleep is further classified into four different stages (NREM 1–4). The EEG of NREM shows high amplitude slow waves reflecting a synchronous hyperpolarization of neurons, whereas the EEG of REM sleep exhibits low voltage, reflecting a lesser degree of synchronization similar to that occurring during wakefulness. These EEG-defined sleep states differ in several behavioral and physiological parameters, such as brain metabolism (Maquet, 2000) and thermoregulation (Parmeggiani, 2003). Sleep is regulated by circadian and homeostatic mechanisms, which are partly independent. The circadian system plays a crucial role in the timing and consolidation of sleep to an ecologically appropriate period; that is, diurnal animals typically sleep during the night and nocturnal animals during the day (Cirelli, 2009). The homeostatic mechanism reflects the need for sleep that accumulates during prolonged periods of wakefulness and dissipates during sustained sleep.

Although sleep is ubiquitous in the animal kingdom, its adaptive value remains elusive. Many explanations have been proposed for sleep function, including energy conservation, restoration at the cellular and network levels, memory consolidation, and synaptic homeostasis (Cirelli and Tononi, 2008; Mignot, 2008). Sleep seems to be particularly important for the brain, because the most immediate effect of sleep deprivation is cognitive impairment (Cirelli and Tononi, 2008). Numerous studies show that sleep benefits or facilitates various brain functions including neural maintenance (Cirelli *et al.*, 2005; Kavanau, 1996; Tononi and Cirelli, 2003, 2006), neurogenesis (Guzman-Marin *et al.*, 2005), and learning and memory (Ambrosini and Giuditta, 2001; Deregnaucourt *et al.*, 2005; Donlea *et al.*, 2011; Huber *et al.*, 2004; Stickgold and Walker, 2005;

Walker and Stickgold, 2004). The recent accumulation of evidence for conservation in the genetics of sleep in vertebrates and invertebrates demonstrates that invertebrates can provide useful model systems for studying the organizational principles and adaptive values of sleep, as well as for deciphering the underlying molecular and neuronal pathways (Allada and Siegel, 2008; Cirelli, 2009).

A. Sleep in invertebrates

The best studied invertebrate model for sleep is the fruit fly *D. melanogaster*. In 2000, two independent research groups provided conclusive evidence that *Drosophila* sleep meets the essential criteria for defining it as sleep (Hendricks *et al.*, 2000; Shaw *et al.*, 2000). Flies show a behavioral quiescence accompanied by an elevated arousal threshold and a homeostatic regulatory mechanism. Pharmacological and molecular studies have revealed additional similarities between sleep physiology in flies and mammals. For example, adenosine antagonists such as caffeine increase wakefulness whereas antihistamines increase rest and reduce its latency (reviewed in Cirelli, 2009; Allada and Siegel, 2008; Crocker and Sehgal, 2010; Greenspan *et al.*, 2001). There is also evidence that sleep and wakefulness in *Drosophila* as well as in other invertebrates are accompanied by characteristic patterns of electrical brain activity (Andretic *et al.*, 2005; Nitz *et al.*, 2002; Ramon *et al.*, 2004; van Swinderen *et al.*, 2004). In *Drosophila*, the local field potentials in the median brain measured during wakefulness differ from those during sleep (Nitz *et al.*, 2002). However, it is not known whether *Drosophila* and other insects also show an increased synchrony of neuronal firing during sleep as is typical of mammals. Fruit flies exhibit an increase in sleep duration (the index for sleep was continuous bouts of 5 min. with no locomotor activity) following sleep deprivation (Hendricks *et al.*, 2000; Shaw *et al.*, 2000). The amount of sleep recovered following sleep deprivation is proportional to the amount of sleep loss. Transgenic manipulation of gene expression in restricted brain areas have begun to characterize the neuronal networks involved in sleep regulation. These studies have identified the mushroom bodies (Joiner *et al.*, 2006; Pitman *et al.*, 2006), the main integration center in the insect brain, and the dorsal fan-shaped body (Donlea *et al.*, 2011), as central for the expression of sleep. Studies on *Drosophila* over the last decade have significantly extended the list of genes involved in various aspects of sleep regulation (reviewed in Cirelli, 2009; Allada and Siegel, 2008; Crocker and Sehgal, 2010).

Another model for sleep research in insects is the honey bee *A. mellifera* (Eban-Rothschild and Bloch, 2012). Honey bees are among the first invertebrates for which a sleep-like state was described (Kaiser and Steiner-Kaiser, 1983). Honey bee foragers exhibit all three behavioral characteristics of sleep: a period of quiescence, an increased response threshold, and a homeostatic regulation mechanism. Foragers sleep in a typical posture characterized by the relaxation of the thorax, head, and antennae, and little, if any, antennae movement (Eban-Rothschild and Bloch, 2008; Kaiser, 1988; Sauer *et al.*, 2003, 2004). They typically sleep in the nest periphery in which the ambient temperature is around 28 °C, similar to their preferred sleep temperature (Kaiser, 1988; Klein *et al.*, 2008; Schmolz *et al.*, 2002). This preference for moderate temperatures probably allows them to conserve energy, and at the same time, it is sufficient for maintaining regenerative processes during sleep (Schmolz *et al.*, 2002). As in other animals, sleep in honey bees is associated with an increase in response threshold. Long-term, extracellular, single-unit recordings from optomotor interneurons in the optic lobes of foragers revealed a diurnal oscillation in their sensitivity to moving visual stimuli; the response threshold was higher during the subjective night than during the subjective day (Kaiser and Steiner-Kaiser, 1983). Elevated response thresholds during sleep were also found for heat and light stimuli (Eban-Rothschild and Bloch, 2008; Kaiser, 1988). For example, the light intensity needed to elicit a response (moving the head and antennae) from a bee in the deepest sleep stage (third sleep stage, see below) is about 10,000 times higher than that needed to obtain a similar response from an immobile awake bee (Eban-Rothschild and Bloch, 2008).

Sleep in honey bees is a dynamic process. Honey bees switch between sleep states that differ from one another in body and antennae posture, bout duration, antenna movements, ventilatory cycle duration, and response threshold (Eban-Rothschild and Bloch, 2008; Sauer *et al.*, 2003). In honey bees as in mammals, the transitions from arousal to deep sleep and from deep sleep to awake states are gradual; there is no evidence, however, for regular sleep cycles with a consistent period as seen in humans and other mammals (Eban-Rothschild and Bloch, 2008). There is a preliminary study suggesting that deep sleep in honey bees is correlated with rhythmic electrophysiological activity in the mushroom bodies (Schuppe, 1995). This is an interesting suggestion because the mushroom bodies are involved in sleep regulation in *Drosophila* (Joiner *et al.*, 2006; Pitman *et al.*, 2006). Honey bees do not show a notable increase in immobility following sleep deprivation, but they do seem to increase the amount of deep sleep (inferred from the analysis of antennal movement; Sauer *et al.*, 2004; Hussaini *et al.*, 2009). The latency from the beginning of the dark period to the first episode of antennal immobility ("sleep latency") also tends to decrease following sleep deprivation (Sauer *et al.*, 2004). This study also showed that disturbing the honey bees during the light period (day) did not result in a similar rebound, suggesting that the response was due to sleep loss and not to the stress associated with the sleep deprivation procedure (Sauer *et al.*, 2004).

In many animals, sleep is associated with memory consolidation, the process that transforms new memories into more stable representations that become integrated into the network of preexisting long-term memories

(Stickgold, 2005). Hussaini *et al.* (2009) found that sleep depriving honey bee foragers had no effect on memory acquisition, but caused a significant reduction in extinction learning in an associative odor learning paradigm. Honey bees that were disturbed during the night showed a slight impairment in the directional, but not distance, precision in their waggle dance communication. A similar impairment was not seen in bees disturbed during the day (Klein *et al.*, 2010). These studies on bees are consistent with studies in mammals in which sleep deprivation impaired performance in some learning/cognitive paradigms but not in others (Stickgold, 2005).

B. Sleep in honey bees with and without circadian rhythms

Honey bees provide a good model for studying plasticity in sleep because their plasticity in circadian rhythms is expected to influence sleep as well (Section III. C above). Because young honey bees are active around the clock with no circadian rhythms, it was unclear whether they sleep as forager bees do. This question was addressed both in the lab, which enabled rigorous quantification of sleep, and in the colony, in which sleep quantification was more challenging, but the context more natural. These complementary approaches showed that honey bees active around the clock do sleep (Eban-Rothschild and Bloch, 2008; Klein *et al.*, 2008). However, in contrast to foragers that exhibit a consolidated nightly sleep (Kaiser, 1988; Sauer *et al.*, 2003), young bees distribute their sleep through-out the day (Eban-Rothschild and Bloch, 2008; Klein *et al.*, 2008). The propor-tion of sleep during the day and the sleep dynamics also differ between honey bees performing different tasks (sleep was defined as a quiescent state with no antennae movements for ∼3 s; Klein *et al.*, 2008). Young bees were more likely to sleep inside cells compared to older ones and, overall, most of the time that they seemed to be sleeping, they were inside cells (Klein *et al.*, 2008). These findings are consistent with previous studies in which the amount of "standing motionless" (a weaker proxy for sleep) was found to be higher for young bees (Moore *et al.*, 1998). Nevertheless, because foragers are also older than nurses, and have a different life experience, additional studies are needed to determine the contribution of social factors, age, and experience to the regulation of plasticity in sleep. Eban-Rothschild and Bloch (2008) video recorded the sleep of individually isolated honey bees in the lab. Young bees (3 days old when transferred from the hive to the lab) showed the same three sleep stages recorded for foragers (typically older than 3 weeks of age), with a similar number of antenna movements and response thresholds for each sleep stage (Eban-Rothschild and Bloch, 2008). As in the colony, the main difference between honey bees with and without circadian rhythms was the distribution of sleep bouts through-out the day; foragers showed consolidated sleep during the subjective night whereas the young bees slept about the same amount of time during the day and night. Young honey bees also differed from foragers in their sleep dynamics; they showed fewer sleep bouts/day overall, but these bouts tended to be longer than those of foragers (Eban-Rothschild and Bloch, 2008). In addition, foragers progressed mainly from light to deep sleep, and from deep sleep they passed directly to awake states, switching less often between sleep stages. Young honey bees tended to pass more often between the three sleep stages without exiting sleep to an awake state. Because the honey bees in this study were removed from the hive, the differences in their sleep behavior cannot be explained by variation in the social environment; they were rather influenced by age or previous experience (including social experience).

V. SOCIAL INFLUENCES ON SLEEP IN INSECTS

The influences of the social environment on sleep were recently reviewed in depth by Donlea and Shaw (2009) and are therefore only briefly summarized here. *Drosophila* flies and other animals experiencing enriched social environments typically show an increase in the total amount of sleep (Bushey *et al.*, 2011; Donlea *et al.*, 2009; Donlea and Shaw, 2009; Donlea *et al.*, 2011; Ganguly-Fitzgerald *et al.*, 2006). For example, flies that were placed with ∼40 siblings for 5 days slept for approximately 2 h more per day than siblings that were socially isolated for 5 days (Ganguly-Fitzgerald *et al.*, 2006). Interestingly, the increase in sleep correlated with the size of the group (i.e., number of flies; Donlea *et al.*, 2009; Donlea *et al.*, 2011). This social experience-dependent increase in sleep was shown to require the ability to form new memories in that a similar increase in sleep was not observed in flies with mutations in genes involved in learning and memory (Ganguly-Fitzgerald *et al.*, 2006).

But why do animals experiencing enriched social environments sleep more? One plausible explanation is that social interactions increase the amount the brain has to invest in the mechanism of neural plasticity, which in turn increases the need for sleep (Donlea and Shaw, 2009). Consistent with this premise, both vertebrates and invertebrates show an elaboration in neural circuitry and synaptic potentiation following a period of enriched social experience. For example, in socially enriched rats there is an increase in the number of dendritic branches in neurons in the visual cortex (Volkmar and Greenoug, 1972), as well as in the number of synapses in the hippocampus (Briones *et al.*, 2006). Similarly, in the *Drosophila* brain, synapse size and number, and the levels of proteins associated with synapse strength increase progressively with the length of time awake (Bushey *et al.*, 2011; Donlea *et al.*, 2009; Gilestro *et al.*, 2009). A plausible explanation for how these wake-dependent changes affect sleep need is suggested by the "*synaptic homeostasis*" hypothesis (Tononi and Cirelli, 2003, 2006). This hypothesis states that strong synapses consume energy, space, and supply, and therefore a continuous increase in synapse strength is unsustainable and could lead to saturation in learning and memory capacity. Sleep, according to this hypothesis, is needed for downscaling synaptic strength to a baseline level that is energetically sustainable (Tononi and Cirelli, 2003, 2006). Consistent with this hypothesis, a decrease in synapse number and space, as well as in the levels of synapse proteins, was recorded after sleep, but not in flies that were sleep deprived for the same period (Bushey *et al.*, 2011; Donlea *et al.*, 2009; Gilestro *et al.*, 2009). This hypothesis further predicts that the more the animal learns and responds to its environment, the more it also needs to sleep. Thus, social enrichment and other intense wake experiences are predicted to increase synaptic strength in various brain circuits, and consequently, the need for sleep. In fact, flies that were housed in large cages in which they could fly and interact with other flies showed an increase in spine and branch number in the unambiguously recognizable VS1 neuron (Bushey *et al.*, 2011). In another study on male flies and a different neuronal circuit, the number of varicosities increased in proportion to the number of flies in the group. As expected from the synaptic homeostasis hypothesis, social enrichment impaired long-term memory in a courtship conditioning paradigm (Donlea *et al.*, 2011). Long-term memory was recovered after 3 days in which the flies could sleep, apparently because this period is sufficient for downscaling the synapses recorded in the study to baseline levels (Donlea *et al.*, 2009).

Besides understanding the mechanisms underlying the experience-dependent increase in sleep, an important challenge is to identify and characterize the social signals that influence this increase. Honey bees provide a good model to address this question because their rich social behavior is well characterized, and can be manipulated in the context of their natural sociobiology. Recent studies suggest that similar to *Drosophila*, honey bees also show an increase in the amount of sleep when experiencing an enriched social environment (Fig. 1.1; Eban-Rothschild and Bloch, in preparation). Young honey bees that experienced the colony environment for their first 24 or 48 h slept more when removed from the hive compared to bees of a similar age and genotype that spent the same period isolated in individual cages. This result is intriguing in that 24 h in the colony was not sufficient time for the development of circadian rhythms (see Fig. 1.1 and Section III.B; Eban-Rothschild *et al.*, 2012), but seemed to influence the amount of sleep. Honey bees that were kept in single- or double-mesh enclosures inside the hive slept the same amount of time as bees moving freely in the hive (Fig. 1.1). The double-mesh enclosures prevented direct contact with other workers, the brood, or the queen. Thus, factors such as volatile pheromones that can permeate the double-mesh enclosures are sufficient to account for the full effect of the colony environment on the amount of sleep. Young bees that spent 2 days outside the hive with 30 additional same-age bees showed intermediate levels of sleep compared to isolated bees and bees that had experienced the colony environment. This finding suggests that social interactions with other bees outside the context of the colony increased the honeybees' need for sleep, but to a less extent than the increase induced by the rich environment of the colony. Again, this effect differs from that concerning circadian rhythms (see Fig. 1.1 and Section III.B). The bees that were transferred from the rich environment of the colony to isolation in the lab showed a decrease in the amount of sleep with time in isolation; after a few days in the lab, these bees slept for the same amount of time as the bees that were individually isolated shortly after emergence. A similar decrease in the amount of sleep was not observed in bees that were individually isolated immediately after emergence and therefore did not experience a decrease in environmental enrichment following the transfer to the monitoring room. A working hypothesis stemming from these findings is that volatile pheromones are sufficient for conveying the complexity of the colony environment and therefore lead to the maximal observed increase in the amount of sleep in bees. Additional studies are needed to test this hypothesis and to identify the social/ environmental signals causing this increase in the amount of sleep.

VI. CONCLUSIONS

Research on social influence on circadian rhythms has traditionally focused on social entrainment in mammals. Our review shows that social influences on the circadian system are richer than being merely entrainment of the phase of activity. For example, in honey bees, social signals not only convincingly entrain circadian rhythms, but also influence their development and context-dependent expression. These influences on the clock appear to be functionally significant. Plasticity in circadian rhythms appears to contribute to task specialization and therefore improves the division of labor and overall colony efficiency (Bloch, 2009, 2010). Similarly, social entrainment appears to facilitate the temporal coordination of individuals interacting in a densely populated and complex society. Social entrainment is also a way to provide important temporal information to individuals that spend most or all of their time in the dark cavity of the nest. For example, temporal social signals can help nectar receivers coordinate their activity with that of nectar foragers, or entrain the clock of young nurses before they perform their first orientation flight (Capaldi et al., 2000).

Studies on Drosophila demonstrate that social influences on sleep and circadian rhythms are not limited to highly social insects, and may be common even in solitary insects. Oscillating pheromonal signals appear to mediate social synchronization in male flies, and the social environment has profound influences on sleep. There is even evidence suggesting social influences on circadian processes other than phase entrainment (see Section III.D). The evidence for potent social influences in Drosophila fit with studies on mammals in which the efficacy of social entrainment is not associated with the degree of sociality. It is intuitive to expect that social influences on the clock would be more important in social animals. The rich repertoire of social influences on the honey bee clock and its apparent functional significance is currently the best support for the hypothesis that social evolution has shaped the circadian system of highly social insects (Bloch, 2009, 2010). However, the studies summarized in Table 1.1 and elsewhere in this paper show that information on the social lifestyle of an animal species is not sufficient for predicting the efficacy of social entrainment. Studies on bats and honey bees suggest that social entrainment is important in dark cavity-dwelling social species. It is therefore possible that the ecology and life history of an animal species are crucial in shaping the sensitivity of its circadian system to social time givers. This hypothesis needs to be tested by comparative studies with additional species.

Little is known about the mechanisms underlying social influences on circadian rhythms. It has been argued that in mammals there are no "social input pathways" linking the sensory systems detecting the social signal to the circadian system. Rather, it was suggested that social entrainment is mediated by non-specific mechanisms such as arousal, social gating of photic input, or motor activity that feeds back to influence the clock (Mistlberger and Skene, 2004). Studies on flies and honey bees highlight the importance of volatile odors and the olfactory system. In flies, there is also evidence that circadian rhythms in odor sensitivity are regulated by peripheral clocks in the olfactory system. It is not clear whether the olfactory signals involved in social entrainment feed into the well-characterized brain circadian network, or act via alternative pathways by which peripheral clocks in the olfactory system regulate locomotor activity. In honey bees, volatile signals appear to be important for social entrainment and for the ontogeny of circadian rhythms, whereas task-related plasticity in circadian rhythms is regulated by direct contact with the brood. These studies show that multiple signals and sensory modalities influence the circadian system of the honey bee.

There is a better conceptual framework for understanding social influences on sleep. Both *Drosophila* flies and honey bees experiencing an enriched social environment sleep more. These findings can be explained by social influences that provide enhanced wake experiences that require enhanced learning, memory, and processing of sensory information. These processes in turn necessitate sleep to support underlying neuronal processes such as memory consolidation or synaptic downscaling. As discussed above, recent studies on *Drosophila* lend credence to the synaptic homeostasis hypothesis. Consistent with this hypothesis, flies in enriched social environments show a progressive elevation in both synaptic strength and need for sleep. It would be worthwhile exploring whether social enrichment influences synaptic strength and sleep need differently compared with enrichment by nonsocial environ-mental signals.

Studies on honey bees have begun to test how different components in the environment influence sleep need. The findings that honey bees with rich wake experiences in the colony sleep more than honey bees caged individually or in a group are consistent with the synaptic homeostasis hypothesis. On the other hand, it is more difficult to reconcile this hypothesis with the findings suggesting that bees caged in single- or double-mesh enclosures slept for the same amount of time as bees moving freely in the hive. This is because the enclosed bees were deprived of many important social signals, and were therefore expected to have a poorer social environment. Perhaps the volatile signals in the hive are sufficiently rich to affect the quantity of sleep in a way comparable to the entire colony environment. Additional studies testing the influence of the social environment on synaptic plasticity and sleep in honey bees are needed.

Social influences on sleep and circadian rhythms in insects have been studied mostly in two species: the fruit fly *D. melanogaster* and the Western honey bee *A. mellifera*. Our review shows that these two model systems are in many senses complementary. The availability of a broad and powerful transgenic toolkit makes *Drosophila* an excellent model with which to study the molecular and neuronal underpinnings governing the interplay between the social environment, sleep, and circadian rhythms. The social behavior of the honey bee can be studied and manipulated in an ecologically natural context. The extensive data on their sociobiology and communication provides means to identify the specific social signals influencing the circadian and sleep regulating systems, and to study the functional significance of these social influences. Recent progress in research on honey bees and flies has created new opportunities for sociomolecular research. The growing interest in *Drosophila* behavior has led to the development of new methods for studying social behavior in this tiny fly. These studies have indicated that the social behavior of *Drosophila* is richer than previously appreciated (reviewed in Sokolowski, 2010). The sequencing of the honey bee genome and the development of functional genomics tools now make it feasible to study the complex social behavior of honey bees at the molecular level (Weinstock *et al.*, 2006). Future research on flies, bees, and additional insect species has the potential to provide new insights on the social and environmental regulation of sleep and circadian rhythms.

Acknowledgments

Research in the authors' laboratory was supported by grants from the Israeli Science Foundation (ISF), US-Israel Binational Foundation (BSF), German-Israeli Foundation (GIF), the National Institute for Psychobiology in Israel (NIPI), the US-Israel Binational Agricultural Research and Development Fund (BARD), and the Joseph H. and Belle R. Braun Senior Lectureship in Life Sciences. The authors thank two anonymous reviewers for their excellent comments on an earlier version of this chapter.

References

Allada, R., and Siegel, J. M. (2008). Unearthing the phylogenetic roots of sleep. *Curr. Biol.* **18,** R670–R679.

Ambrosini, M. V., and Giuditta, A. (2001). Learning and sleep: The sequential hypothesis. *Sleep Med. Rev.* **5,** 477–490.

Andretic, R., van Swinderen, B., and Greenspan, R. J. (2005). Dopaminergic modulation of arousal in Drosophila. *Curr. Biol.* **15,** 1165–1175.

Beaver, L. M., and Giebultowicz, J. M. (2004). Regulation of copulation duration by period and timeless in Drosophila melanogaster. *Curr. Biol.* **14,** 1492–1497.

Bell-Pedersen, D., Cassone, V. M., Earnest, D. J., Golden, S. S., Hardin, P. E., Thomas, T. L., and Zoran, M. J. (2005). Circadian rhythms from multiple oscillators: Lessons from diverse organisms. *Nat. Rev. Genet.* **6,** 544–556.

Bloch, G. (2009). Plasticity in the circadian clock and the temporal organization of insect societies. *In* "Organization of Insect Societies: From Genome to Sociocomplexity" (J. Gadau and J. Fewell, eds.), pp. 402–432. Harvard University Press, Cambridge.

Bloch, G. (2010). The social clock of the honeybee. *J. Biol. Rhythms* **25,** 307–317.

Bloch, G., and Meshi, A. (2007). Influences of octopamine and juvenile hormone on locomotor behavior and period gene expression in the honeybee, Apis mellifera. *J. Comp. Physiol. A* **193,** 181–199.

Bloch, G., and Robinson, G. E. (2001). Chronobiology—Reversal of honeybee behavioural rhythms. *Nature* **410,** 1048.

Bloch, G., Toma, D. P., and Robinson, G. E. (2001). Behavioral rhythmicity, age, division of labor and period expression in the honey bee brain. *J. Biol. Rhythms* **16,** 444–456.

Bloch, G., Solomon, S. M., Robinson, G. E., and Fahrbach, S. E. (2003). Patterns of PERIOD and pigment-dispersing hormone immunoreactivity in the brain of the European honeybee (Apis mellifera): Age- and time-related plasticity. *J. Comp. Neurol.* **464,** 269–284.

Bloch, G., Rubinstein, C. D., and Robinson, G. E. (2004). period expression in the honey bee brain is developmentally regulated and not affected by light, flight experience, or colony type. *Insect Biochem. Mol. Biol.* **34,** 879–891.

Briones, T. L., Suh, E., Jozsa, L., and Woods, J. (2006). Behaviorally induced synaptogenesis and dendritic growth in the hippocampal region following transient global cerebral ischemia are accompanied by improvement in spatial learning. *Exp. Neurol.* **198,** 530–538.

Bushey, D., Tononi, G., and Cirelli, C. (2011). Sleep and synaptic homeostasis: Structural evidence in Drosophila. *Science* **332,** 1576–1581.

Capaldi, E. A., Smith, A. D., Osborne, J., Fahrbach, S. E., Farris, S. M., Reynolds, D. R., Edwards, A. S., Martin, A., Robinson, G. E., Poppy, G. M., and Riley, J. R. (2000). Ontogeny of orientation flight in the honeybee revealed by harmonic radar. *Nature* **403,** 537–540.

Chatterjee, A., Tanoue, S., Houl, J. H., and Hardin, P. E. (2010). Regulation of gustatory physiology and appetitive behavior by the Drosophila circadian clock. *Curr. Biol.* **20,** 300–309.

Cirelli, C. (2009). The genetic and molecular regulation of sleep: From fruit flies to humans. *Nat. Rev. Neurosci.* **10,** 549–560.

Cirelli, C., and Tononi, G. (2008). Is sleep essential? *PLoS Biol.* **6,** 1605–1611.

Cirelli, C., Furaguna, U., and Tononi, G. (2005). Molecular correlates of long-term sleep deprivation in rats: A genome-wide analysis. *Sleep* **28,** 1000.

Crailsheim, K., Hrassnigg, N., and Stabentheiner, A. (1996). Diurnal behavioural differences in forager and nurse honey bees (*Apis mellifera carnica Pollm*). *Apidologie* **27,** 235–244.

Crocker, A., and Sehgal, A. (2010). Genetic analysis of sleep. *Genes Dev.* **24,** 1220–1235.

Davidson, A. J., and Menaker, M. (2003). Birds of a feather clock together—Sometimes: Social synchronization of circadian rhythms. *Curr. Opin. Neurobiol.* **13,** 765–769.

Deregnaucourt, S., Mitra, P. P., Feher, O., Pytte, C., and Tchernichovski, O. (2005). How sleep affects the developmental learning of bird song. *Nature* **433,** 710–716.

Donlea, J. M., and Shaw, P. J. (2009). Sleeping together: Using social interactions to understand the role of sleep in plasticity. *Adv. Genet.* **68,** 57–81.

Donlea, J. M., Ramanan, N., and Shaw, P. J. (2009). Use-dependent plasticity in clock neurons regulates sleep need in Drosophila. *Science* **324,** 105–108.

Donlea, J. M., Thimgan, M. S., Suzuki, Y., Gottschalk, L., and Shaw, P. J. (2011). Inducing sleep by remote control facilitates memory consolidation in Drosophila. *Science* **332,** 1571–1576.

Dunlap, J. C., Loros, J. J., and Decoursey, P. J. (2004). Chronobiology: Biological Timekeeping. pp. i–xx. Sinauer, Sunderland, MA. 1–406.

Eban-Rothschild, A. D., and Bloch, G. (2008). Differences in the sleep architecture of forager and young honeybees (Apis mellifera). *J. Exp. Biol.* **211,** 2408–2416.

Eban-Rothschild, A., and Bloch, G. (2012). Circadain rhythms and sleep in honey bees. In "Honeybee Neurobiology and Behavior, A Tribute to Randolf Menzel" (G. Galizia, D. Eisenhardt, and M. Giurfa, eds.), pp. 31–46. Springer, Dordrecht/Heidelberg/London/New York.

Eban-Rothschild, A., Shemesh, Y., and Bloch, G. (2012). The colony environment, but not direct contact with conspecifics, influences the development of circadian rhythms in honey bees. *J. Biol. Rhythms* (in press).

Eban-Rothschild, A., and Bloch, G. Social enrichment increases sleep amount in young honey bee (*Apis mellifera*) workers (in preparation).

Eban-Rothschild, A., and Bloch, G. Social synchronization of locomotor activity rhythms in young honey bee (*Apis mellifera*) workers (in preparation).

Eban-Rothschild, A., Belluci, S., and Bloch, G. (2011). Maternity-related plasticity in circadian rhythms of bumble-bee queens. *Proc. R. Soc. Lond. Biol. Sci.* **278,** 3510–3516.

Erkert, H. G., Nagel, B., and Stephani, I. (1986). Light and social effects on the free-running circadian activity rhythm in common marmosets (*Callithrix jacchus*; primates): social masking, pseudo-splitting, and relative coordination). *Behav. Ecol. Sociobiol.* **18,** 444–452.

Frisch, B., and Aschoff, J. (1987). Circadian rhythms in honeybees: Entrainment by feeding cycles. *Physiol. Entomol.* **12,** 41–49.

Frisch, B., and Koeniger, N. (1994). Social synchronization of the activity rhythms of honeybees within a colony. *Behav. Ecol. Sociobiol.* **35,** 91–98.

Fujii, S., and Amrein, H. (2010). Ventral lateral and DN1 clock neurons mediate distinct properties of male sex drive rhythm in Drosophila. *Proc. Natl. Acad. Sci. U.S.A.* **107,** 10590–10595.

Fujii, S., Krishnan, P., Hardin, P., and Amrein, H. (2007). Nocturnal male sex drive in Drosophila. *Curr. Biol.* **17,** 244–251.

Ganguly-Fitzgerald, I., Donlea, J., and Shaw, P. J. (2006). Waking experience affects sleep need in Drosophila. *Science* **313,** 1775–1781.

Gattermann, R., and Weinandy, R. (1997). Lack of social entrainment of circadian activity rhythms in the solitary golden hamster and in the highly social Mongolian gerbil. *Biol. Rhythm Res.* **28,** 85–93.

Gilestro, G. F., Tononi, G., and Cirelli, C. (2009). Widespread changes in synaptic markers as a function of sleep and wakefulness in Drosophila. *Science* **324,** 109–112.

Goel, N., and Lee, T. M. (1997). Olfactory bulbectomy impedes social but not photic reentrainment of circadian rhythms in female Octodon degus. *J. Biol. Rhythms* **12,** 362–370.

Goel, N., and Lee, T. M. (1995). Sex differences and effects of social cues on daily rhythms following phase advances in *Octodon degus*. *Physiol. Behav.* **58,** 205–213.

Greenspan, R. J., Tononi, G., Cirelli, C., and Shaw, P. J. (2001). Sleep and the fruit fly. *Trends Neurosci.* **24,** 142–145.

Guzman-Marin, R., Suntsova, N., Methippara, M., Greiffenstein, R., Szymusiak, R., and McGinty, D. (2005). Sleep deprivation suppresses neurogenesis in the adult hippocampus of rats. *Eur. J. Neurosci.* **22**, 2111–2116.

Hardelan, R. (1972). Species differences in the diurnal rhythmicity of courtship behaviour within the *Melanogaster* group of the genus *Drosophila. Anim. Behav.* **20**, 170.

Helfrich-Foerster, C., Shafer, O. T., Wuelbeck, C., Grieshaber, E., Rieger, D., and Taghert, P. (2007). Development and morphology of the clock-gene-expressing lateral neurons of Drosophila melanogaster. *J. Comp. Neurol.* **500**, 47–70.

Helfrich-Forster, C., Stengl, M., and Homberg, U. (1998). Organization of the circadian system in insects. *Chronobiol. Int.* **15**, 567–594.

Hendricks, J. C., Finn, S. M., Panckeri, K. A., Chavkin, J., Williams, J. A., Sehgal, A., and Pack, A. I. (2000). Rest in Drosophila is a sleep-like state. *Neuron* **25**, 129–138.

Huang, Z. Y., and Otis, G. W. (1991). Inspection and feeding of larvae by worker honey bees (Hymenoptera: Aphidae): Effects of starvation and food quantity. *J. Insect Behav.* **4**, 305–317.

Huber, R., Ghilardi, M., Massimini, M., and Tononi, G. (2004). Local increase in slow-wave activity after a learning task. *Sleep* **27**, 067.

Hussaini, S. A., Bogusch, L., Landgraf, T., and Menzel, R. (2009). Sleep deprivation affects extinction but not acquisition memory in honeybees. *Learn. Mem.* **16**, 698–705.

Ingram, K., Krummey, S., and LeRoux, M. (2009). Expression patterns of a circadian clock gene are associated with age-related polyethism in harvester ants, Pogonomyrmex occidentalis. *BMC Ecol.* **9**, 7.

Joiner, W. J., Crocker, A., White, B. H., and Sehgal, A. (2006). Sleep in Drosophila is regulated by adult mushroom bodies. *Nature* **441**, 757–760.

Jong, J.-J., and Lee, H.-J. (2008). Differential expression of circadian locomotor rhythms among castes of the gray-black spiny ant, Polyrhachis dives (Hymenoptera: Formicidae). *Sociobiology* **52**, 167–184.

Kaiser, W. (1988). Busy bees need rest, too: Behavioral and electromyographical sleep signs in honeybees. *J. Comp. Physiol. A* **163**, 565–584.

Kaiser, W., and Steiner-Kaiser, J. (1983). Neuronal correlates of sleep, wakefulness and arousal in a diurnal insect. *Nature* **301**, 707–709.

Kavanau, J. L. (1996). Memory, sleep, and dynamic stabilization of neural circuitry: Evolutionary perspectives. *Neurosci. Biobehav. Rev.* **20**, 289–311.

Kefuss, J., and Nye, W. (1970). The influence of photoperiod on the flight activity of honeybees. *J. Apic. Res.* **9**, 133–139.

Klein, B. A., Olzsowy, K. M., Klein, A., Saunders, K. M., and Seeley, T. D. (2008). Caste-dependent sleep of worker honey bees. *J. Exp. Biol.* **211**, 3028–3040.

Klein, B. A., Klein, A., Wray, M. K., Mueller, U. G., and Seeley, T. D. (2010). Sleep deprivation impairs precision of waggle dance signaling in honey bees. *Proc. Natl. Acad. Sci. U.S.A.* **107**, 22705–22709.

Kleinknecht, S. (1985). Lack of social entrainment of free-running circadian activity rhythms in the Australian sugar glider (*Petaurus breviceps*: Marsupialia). *Behav. Ecol. Sociobiol.* **16**, 189–193.

Knadler, J. J., and Page, T. I. (2009). Social interactions and the circadian rhythm in locomotor activity in the cockroach Leucophaea maderae. *Chronobiol. Int.* **26**, 415–429.

Koeniger, N., and Koeniger, G. (2000). Reproductive isolation among species of the genus Apis. *Apidologie* **31**, 313–339.

Kostal, V., Zavodska, R., and Denlinger, D. (2009). Clock genes period and timeless are rhythmically expressed in brains of newly hatched, photosensitive larvae of the fly, Sarcophaga crassipalpis. *J. Insect Physiol.* **55**, 408–414.

Krishnan, B., Dryer, S. E., and Hardin, P. E. (1999). Circadian rhythms in olfactory responses of Drosophila melanogaster. *Nature* **400**, 375–378.

Krupp, J. J., Kent, C., Billeter, J. C., Azanchi, R., So, A. K. C., Schonfeld, J. A., Smith, B. P., Lucas, C., and Levine, J. D. (2008). Social experience modifies pheromone expression and mating behavior in male Drosophila melanogaster. *Curr. Biol.* **18**, 1373–1383.

Lee, G., Bahn, J. H., and Park, J. H. (2006). Sex- and clock-controlled expression of the neuropeptide F gene in Drosophila. *Proc. Natl. Acad. Sci.* **103**, 12580–12585.

Levine, J. D., Funes, P., Dowse, H. B., and Hall, J. C. (2002). Resetting the circadian clock by social experience in Drosophila melanogaster. *Science* **298**, 2010–2012.

Lindauer, M. (1961). Communication Among Social Bees. Harvard University Press, Cambridge, MA.

Marimuthu, G., Rajan, S., and Chandrashekaran, M. K. (1981). Social entrainment of the circadian rhythm in the flight activity of the Microchiropteran bat *Hipposideros speoris*. *Behav. Ecol. Sociobiol.* **8**, 147–150.

Maquet, P. (2000). Functional neuroimaging of normal human sleep by positron emission tomography. *J. Sleep Res.* **9**, 207–231.

Meshi, A., and Bloch, G. (2007). Monitoring circadian rhythms of individual honey bees in a social environment reveals social influences on postembryonic ontogeny of activity rhythms. *J. Biol. Rhythms* **22**, 343–355.

Mignot, E. (2008). Why we sleep: The temporal organization of recovery. *PLoS Biol.* **6**, 661–669.

Mistlberger, R. E., and Skene, D. J. (2004). Social influences on mammalian circadian rhythms: Animal and human studies. *Biol. Rev.* **79**, 533–556.

Moore, D. (2001). Honey bee circadian clocks: Behavioral control from individual workers to whole-colony rhythms. *J. Insect Physiol.* **47**, 843–857.

Moore, D., Angel, J. E., Cheeseman, I. M., Fahrbach, S. E., and Robinson, G. E. (1998). Timekeeping in the honey bee colony: Integration of circadian rhythms and division of labor. *Behav. Ecol. Sociobiol.* **43**, 147–160.

Moritz, R. F. A., and Sakofski, F. (1991). The role of the queen in circadian rhythms of honeybees (Apis mellifera L.). *Behav. Ecol. Sociobiol.* **29**, 361–365.

Moritz, R., and Kryger, P. (1994). Self-organization of circadian rhythms in groups of honeybees (*Apis mellifera* L.). *Behav. Ecol. Sociobiol.* **34**, 211–215.

Mrosovsky, N. (1988). Phase response curves for social entrainment. *J. Comp. Physiol.* A **162**, 35–46.

Nagari, M., and Bloch, G. (2012). The involvement of the antennae in mediating the brood influence on circadian rhythms in "nurse" honey bee (*Apis mellifera*) workers. *J Insect Physiol*, (available online). DOI: 10.1016/j.jinsphys.2012.05.007.

Nitabach, M. N., and Taghert, P. H. (2008). Organization of the Drosophila circadian control circuit. *Curr. Biol.* **18**, R84–R93.

Nitz, D. A., van Swinderen, B., Tononi, G., and Greenspan, R. J. (2002). Electrophysiological correlates of rest and activity in Drosophila melanogaster. *Curr. Biol.* **12**, 1934–1940.

Parmeggiani, P. L. (2003). Thermoregulation and sleep. *Front. Biosci.* **8**, S557–S567.

Peschel, N., and Helfrich-Forster, C. (2011). Setting the clock—By nature: Circadian rhythm in the fruitfly Drosophila melanogaster. *FEBS Lett.* **585**, 1435–1442.

Pitman, J. L., McGill, J. J., Keegan, K. P., and Allada, R. (2006). A dynamic role for the mushroom bodies in promoting sleep in Drosophila. *Nature* **441**, 753–756.

Prober, D. A., Rihel, J., Onah, A. A., Sung, R. J., and Schier, A. F. (2006). Hypocretin/orexin overexpression induces an insomnia-like phenotype in zebrafish. *J. Neurosci.* **26**, 13400–13410.

Raizen, D. M., Zimmerman, J. E., Maycock, M. H., Ta, U. D., You, Y. J., Sundaram, M. V., and Pack, A. I. (2008). Lethargus is a Caenorhabditis elegans sleep-like state. *Nature* **451**, 569–572.

Rajaratnam, S. M. W., and Redman, J. (1999). Social contact synchronizes free-running activity rhythms of diurnal palm squirrels. *Physiol. Behav.* **66**, 21–26.

Ramon, F., Hernandez-Falcon, J., Nguyen, B., and Bullock, T. H. (2004). Slow wave sleep in crayfish. *Proc. Natl. Acad. Sci. U.S.A.* **101**, 11857–11861.

Refinetti, R., Nelson, D. E., and Menaker, M. J. (1992). Social stimuli fail to act as entraining agents of circadian rhythms in the golden hamster. *J. Comp. Physiol.* **170,** 181–187.

Rivkees, S. A. (2003). Developing circadian rhythmicity in infants. *Pediatrics* **112,** 373–381.

Rodriguez-Zas, S. L., Southey, B. R., Shemesh, Y., Rubin, E. B., Cohen, M., Robinson, G. E., and Bloch, G. (2012). Microarray analysis of natural socially regulated plasticity in circadian rhythms of honey bees. *J. Biol. Rhythms* **27,** 12–24.

Rubin, E. B., Shemesh, Y., Cohen, M., Elgavish, S., Robertson, H. M., and Bloch, G. (2006). Molecular and phylogenetic analyses reveal mammalian-like clockwork in the honey bee (Apis mellifera) and shed new light on the molecular evolution of the circadian clock. *Genome Res.* **16,** 1352–1365.

Sakai, T., and Ishida, N. (2001). Circadian rhythms of female mating activity governed by clock genes in Drosophila. *Proc. Natl. Acad. Sci. U.S.A.* **98,** 9221–9225.

Sasagawa, H., Narita, R., Kitagawa, Y., and Kadowaki, T. (2003). The expression of genes encoding visual components is regulated by a circadian clock, light environment and age in the honeybee (Apis mellifera). *Eur. J. Neurosci.* **17,** 963–970.

Sasaki, M. (1990). Photoperiodic regulation of honeybee mating-flight time: Exploitation of innately phase-fixed circadian oscillation. *Adv. Invert. Reprod.* **5,** 503–508.

Sauer, S., Kinkelin, M., Herrmann, E., and Kaiser, W. (2003). The dynamics of sleep-like behaviour in honey bees. *J. Comp. Physiol. A Neuroethol. Sens. Neural Behav. Physiol.* **189,** 599–607.

Sauer, S., Herrmann, E., and Kaiser, W. (2004). Sleep deprivation in honey bees. *J. Sleep Res.* **13,** 145–152.

Saunders, D. S. (2002). Insect Clocks. Elsevier Press, Amsterdam.

Schmolz, E., Hoffmeister, D., and Lamprecht, I. (2002). Calorimetric investigations on metabolic rates and thermoregulation of sleeping honeybees (Apis mellifera carnica). *Thermochim. Acta* **382,** 221–227.

Schuppe, H. (1995). Rhythmic brain activity in sleeping bees. *Wien. Med. Wochenschr.* **145,** 463–464.

Sehadova, H., Sauman, I., and Sehnal, F. (2003). Immunocytochemical distribution of pigment-dispersing hormone in the cephalic ganglia of polyneopteran insects. *Cell Tissue Res.* **312,** 113–125.

Sehadova, H., Markova, E. P., Sehnal, F. S., and Takeda, M. (2004). Distribution of circadian clock-related proteins in the cephalic nervous system of the silkworm, Bombyx mori. *J. Biol. Rhythms* **19,** 466–482.

Sehgal, A., and Mignot, E. (2011). Genetics of sleep and sleep disorders. *Cell* **146,** 194–207.

Sehgal, A., Price, J., and Young, M. W. (1992). Ontogeny of a biological clock in *Drosophila melanogaster. Proc. Natl. Acad. Sci. U.S.A.* **89,** 1423–1427.

Shaw, P. J., Cirelli, C., Greenspan, R. J., and Tononi, G. (2000). Correlates of sleep and waking in Drosophila melanogaster. *Science* **287,** 1834–1837.

Shemesh, Y., Cohen, M., and Bloch, G. (2007). Natural plasticity in circadian rhythms is mediated by reorganization in the molecular clockwork in honeybees. *FASEB J.* **21,** 2304–2311.

Shemesh, Y., Eban-Rothschild, A., Cohen, M., and Bloch, G. (2010). Molecular dynamics and social regulation of context-dependent plasticity in the circadian clockwork of the honey bee. *J. Neurosci.* **30,** 12517–12525.

Siegel, J. M. (2005). Clues to the functions of mammalian sleep. *Nature* **437,** 1264–1271.

Sokolowski, M. B. (2010). Social interactions in "simple" model systems. *Neuron* **65,** 780–794.

Southwick, E. E., and Moritz, R. F. A. (1987). Social synchronization of circadian rhythms of metabolism in honeybees (Apis mellifera). *Physiol. Entomol.* **12,** 209–212.

Spangler, H. G. (1972). Daily activity rhythms of individual worker and drone honeybees. *Ann. Entomol. Soc. Am.* **65,** 1073–1076.

Stickgold, R. (2005). Sleep-dependent memory consolidation. *Nature* **437,** 1272–1278.

Stickgold, R., and Walker, M. P. (2005). Memory consolidation and reconsolidation: What is the role of sleep? *Trends Neurosci.* **28,** 408–415.

Stussi, T., and Harmelin, M. L. (1966). Recherche sur l'ontogenese du rythme circadien de la depense d'energie chez l'Abeille. *C R Acad. Sci. Hebd. Seances Acad. Sci. D* **262**, 2066–2069.

Tauber, E., Roe, H., Costa, R., Hennessy, J. M., and Kyriacou, C. P. (2003). Temporal mating isolation driven a behavioral gene in Drosophila. *Curr. Biol.* **13**, 140–145.

Tobler, I. (1983). Effect of forced locomotion on the rest-activity cycle of the cockroach. *Behav. Brain Res.* **8**, 351–360.

Toma, D. P., Bloch, G., Moore, D., and Robinson, G. E. (2000). Changes in period mRNA levels in the brain and division of labor in honey bee colonies. *Proc. Natl. Acad. Sci. U.S.A.* **97**, 6914–6919.

Tononi, G., and Cirelli, C. (2003). Sleep and synaptic homeostasis: A hypothesis. *Brain Res. Bull.* **62**, 143–150.

Tononi, G., and Cirelli, C. (2006). Steep function and synaptic homeostasis. *Sleep Med. Rev.* **10**, 49–62.

van Swinderen, B., Nitz, D. A., and Greenspan, R. J. (2004). Uncoupling of brain activity from movement defines arousal states in Drosophila. *Curr. Biol.* **14**, 81–87.

Vanlalnghaka, C., and Joshi, D. S. (2005). Entrainment by different environmental stimuli in the frugivorous bats from the Lonar crater. *Biol. Rhythm Res.* **36**, 445–452.

Volkmar, F. R., and Greenoug, W. T. (1972). Rearing complexity affects branching of dendrites in the visual cortex of the rat. *Science* **176**, 1445–1447.

Walker, M. P., and Stickgold, R. (2004). Sleep-dependent learning and memory consolidation. *Neuron* **44**, 121–133.

Weinstock, G. M., Robinson, G. E., Gibbs, R. A., Worley, K. C., Evans, J. D., Maleszka, R., Robertson, H. M., Weaver, D. B., Beye, M., Bork, P., Elsik, C. G., Hartfelder, K., *et al.* (2006). Insights into social insects from the genome of the honeybee Apis mellifera. *Nature* **443**, 931–949.

Wilson, E. O. (1971). The Insect Societies. Belknap Press of Harvard University Press, Cambridge, MA.

Winston, M. L. (1987). The Biology of the Honey Bee. Harvard University Press, Cambridge, MA.

Yellin, A. M., and Hauty, G. T. (1971). Activity cycles of the rhesus monkey (*Macaco mulatta*) under several experimental conditions, both in isolation and in a group situation. *J. Interdisciplinary Cycle Res.* **2**, 475–490.

Yerushalmi, S., Bodenhaimer, S., and Bloch, G. (2006). Developmentally determined attenuation in circadian rhythms links chronobiology to social organization in bees. *J. Exp. Biol.* **209**, 1044–1051.

Yuan, Q., Metterville, D., Briscoe, A. D., and Reppert, S. M. (2007). Insect cryptochromes: Gene duplication and loss define diverse ways to construct insect circadian clocks. *Mol. Biol. Evol.* **24**, 948–955.

Zavodska, R., Sauman, I., and Sehnal, F. (2003). Distribution of PER protein, pigment-dispersing hormone, prothoracicotropic hormone, and eclosion hormone in the cephalic nervous system of insects. *J. Biol. Rhythms* **18**, 106–122.

Zhang, E. E., and Kay, S. A. (2010). Clocks not winding down: Unravelling circadian networks. *Nat. Rev. Mol. Cell Biol.* **11**, 764–776.

2 | Interplay Between Social Experiences and the Genome: Epigenetic Consequences for Behavior

Frances A. Champagne

Department of Psychology, Columbia University, New York, USA

ABSTRACT

Social experiences can have a persistent effect on biological processes leading to phenotypic diversity. Variation in gene regulation has emerged as a mechanism through which the interplay between DNA and environments leads to the biological encoding of these experiences. Epigenetic modifications—molecular pathways through which transcription is altered without altering the underlying DNA sequence—play a critical role in the normal process of development and are being increasingly explored as a mechanism linking environmental experiences

Advances in Genetics, Vol. 77
0065-2660/12 $35.00
http://dx.doi.org/10.1016/B978-0-12-387687-4.00002-7

to long-term biobehavioral outcomes. In this review, evidence implicating epigenetic factors, such as DNA methylation and histone modifications, in the link between social experiences occurring during the postnatal period and in adulthood and altered neuroendocrine and behavioral outcomes will be highlighted. In addition, the role of epigenetic mechanisms in shaping variation in social behavior and the implications of epigenetics for our understanding of the transmission of traits across generations will be discussed.   © 2012, Elsevier Inc.

Though our DNA sequence provides a template for generating individual differences and for the transmission of those characteristics across generations, it is evident that the environment contributes significantly to both of these processes. The dichotomy between the influences of nature (DNA) and nurture (environment) has become part of our theoretical framework for understanding the origins of our unique characteristics—ranging from personality to disease risk. However, advances in molecular biology have provided enlightenment regarding both the relationship between genotype and phenotype and the mechanisms through which environment shapes biological processes, and the insights gained from these advances has challenged the nature versus nurture dichotomy. It is becoming increasingly apparent that development is a process of dynamic interplay between the genome and our environmental experiences. This interplay is perhaps best illustrated in the concept of epigenetics—the study of those factors which alter the activity of genes without altering the underlying DNA sequence. There is increasing evidence that environmental experiences can come to shape the activity of genes through epigenetic pathways and thus the genotype to phenotype and environment to phenotype relationships may ultimately be governed by similar molecular processes.

The investigation of environmentally induced epigenetic effects has highlighted the impact of a wide variety of environmental exposures, including nutritional levels during development, toxins, and both early and later life stressors (Champagne, 2010; Jirtle and Skinner, 2007). However, it is perhaps the interplay between social experiences and the genome which demonstrates most profoundly how nature and nurture interact. Across a variety of species, there is evidence for the effect of social experiences occurring across the lifespan on epigenetic pathways leading to broad phenotypic effects, including stress responsivity, learning/memory, and reproductive behavior (Champagne, 2010). Moreover, epigenetic mechanisms may account for the emergence of social experiences through the effects of these molecular modifications on social behavior (Auger et al., 2011; Kataoka et al., 2011). Ultimately, these interactive effects can lead to the transgenerational continuity of individual differences in neurobiological and behavioral characteristics (Champagne, 2008). In this chapter, evidence for the epigenetic consequence of social experiences, such as mother–infant interactions and adult social interactions, will be highlighted followed by a discussion of the epigenetic basis of variation in social behavior.

This research contributes to our understanding of the inheritance of behavior and raises challenging questions regarding future directions in the study of behavioral epigenetics.

I. EPIGENETIC MECHANISMS AND DEVELOPMENT

The regulation of gene transcription is a dynamic process and illustrates the complexity of the genotype to phenotype relationship. Epigenetic modifications are a key component of this process and can act to enhance or suppress transcription through a variety of pathways. The term "epigenetic" has historically been used to describe the interplay between genes and gene products which can result in variation in the phenotype of an organism (Jablonka and Lamb, 2002; Waddington, 1942). The term predates modern approaches to genetics and suggests that there are factors "over" or "upon" genetic variation that must be considered within the study of developmental biology. However, more modern uses of the term are in the description of the specific molecular mechanisms that, through interactions with DNA or DNA products (mRNA), lead to variation in gene expression. Though our understanding of the complexity and variety of these mechanisms is rapidly expanding, here the focus will be on epigenetic factors that have been studied in the context of developmental plasticity in behavior in response to the environment and variation in social behavior: DNA methylation, histone modifications, and microRNAs (miRNAs).

A. DNA methylation

The accessibility of DNA to transcription factors and RNA polymerase is a critical step in the initiation of transcription. DNA methylation is a molecular modification which alters this accessibility without altering the underlying DNA sequence (Turner, 2001). Cytosine nucleotides within the DNA sequence can become methylated through the chemical addition of a methyl group to the 5 position of the cytosine pyrimidine ring (see Fig. 2.1A). This modification is not a mutation—the cytosine can still make an appropriate base pairing with guanine—however, the cytosine becomes less accessible to transcription factors. Moreover, methylated DNA attracts methyl-binding proteins (MBDs), such as methyl-CpG-binding protein 2 (MeCP2), which can further modify the chromatin structure to reduce transcription (Fan and Hutnick, 2005; Razin, 1998). DNA methylation typically occurs at CpG dinucleotides, which are often found in gene promoters (i.e., DNA sequences upstream from the transcription start site which regulate gene activity) (Deaton and Bird, 2011). The addition of methyl groups to cytosines is accomplished through the actions of the DNA methyltransferases (DNMTs). Within this class of enzymes are the maintenance

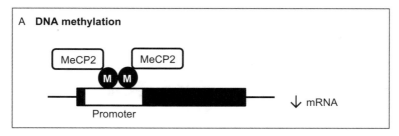

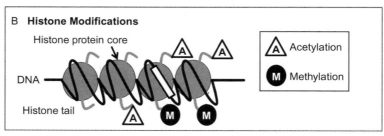

Figure 2.1. Schematic illustration of DNA methylation and histone modifications. (A) DNA methylation involves the formation of a chemical bond between a methyl group (M) and cytosines within the DNA sequence. Methylated cytosines attract methyl-binding proteins such as MeCP2. When cytosine methylation occurs within the promoter region of a gene, there is typically a reduction in transcriptional activity of the gene leading to reduced mRNA levels. (B) DNA is wrapped around a core of histone proteins. The amino acid tails of these proteins can undergo multiple posttranslational modifications, including acetylation (A) and methylation (M). These modifications may alter the interactions between histone tails and the surrounding DNA sequence resulting in altered gene transcription.

DNMTs—such as DNMT1—which methylate hemimethylated DNA following DNA replication as well as the *de novo* methyltransferases (DNMT3a/b) which add methyl groups to unmethylated DNA (Okano *et al.*, 1999; Turek-Plewa and Jagodzinski, 2005). The mechanism of removal of methyl marks within DNA has been far more elusive, particularly in light of the strong covalent bond which characterizes this chemical modification. However, during DNA replication and repair, there are opportunities for the passive loss of DNA methylation. Drugs that inhibit DNMT1—such as 5-azacytidine—have been demonstrated to reduce DNA methylation levels, oxidation of 5-methylcytosine to 5-hydroxymethylcytosine may promote subsequent demethylation, and recent evidence suggests there may be specific enzymes with demethylase properties which can induce DNA demethylation in postmitotic cells (Nabel and Kohli, 2011; Wu and Zhang, 2010). Thus, there is the capacity to both add and remove methyl groups to/from DNA with implications for transcriptional activity.

DNA methylation is a fundamental process within development and functions to maintain patterns of cellular differentiation. This function is accomplished via two important properties of DNA methylation—stability and heritability. Though there are processes which can lead to demethylation and genome-wide demethylation is observed in most species during early embryonic development (Santos *et al.*, 2002), DNA methylation is considered a mechanism of stable gene silencing (Razin, 1998). Moreover, DNA methylation patterns can be maintained following mitosis, thus permitting stability of cell type-specific gene expression profiles (Jones and Taylor, 1980). During the rapid phases of cell proliferation and differentiation that occur during early embryonic development, the process of DNA methylation is essential for survival. Targeted deletion of *DNMT1* or *DNMT3a/b* in mice has been found to be lethal and results in abnormal gene expression patterns (Li *et al.*, 1992; Okano *et al.*, 1999).

B. Histone modifications

The compact storage of DNA within the cell nucleus is accomplished through the condensation of chromatin via the tight packaging of nucleosomes, which are the fundamental units of chromatin and consist of DNA wrapped around a protein core. The protein core of the nucleosome is made up of the histone proteins H2a, H2b, H3, and H4 (Turner, 2001). The N-terminal "tails" of histone proteins can undergo multiple posttranslational modifications which can alter chromatin structure (see Fig. 2.1B). These modifications typically occur at lysine residues throughout the tail and include acetylation, methylation, ubiquitination, and phosphorylation (Jenuwein and Allis, 2001; Peterson and Laniel, 2004). The nature and the location of the modification will determine the effect of this epigenetic modification for gene transcription. For example, acetylation of lysines within the H3 histone is typically associated with increased transcriptional activity, whereas H3 methylation can reduce transcriptional activity (e.g., H3K9 methylation) or be associated with active gene expression (e.g., H3K4 methylation; Barski *et al.*, 2007; Koch *et al.* 2007). The availability of enzymes to facilitate these posttranslational modifications is a key regulatory step in histone-mediated effects (Legube and Trouche, 2003). In the case of acetylation, the histone acetylases promote the occurrence of this modification, whereas histone deacetylases (HDACs) remove the acetyl groups from the histone tails. Similarly, there are enzymes that promote histone methylation (i.e., histone methyltransferases), and enzymes that remove the methyl groups from histone proteins (i.e., histone demethylases). There are multiple types of HDACs (e.g., HDAC1, HDAC5) and histone methyltransferases/demethylases (e.g., EHMT1, JARID1C), defined by tissue and substrate specificity, and mutation of the genes encoding these enzymes can have a profound effect on development (Dokmanovic *et al.*, 2007). Pharmacological manipulation of histone acetylation can be achieved through use of valproic acid, sodium butyrate, and trichostatin A—compounds that inhibit HDACs (Monneret, 2005). The modification of histones is dynamic and can lead to transient and reversible effects on transcription.

C. MicroRNA

miRNAs are small, noncoding RNA molecules which can interfere with mRNA molecules and thus act to modify gene activity through posttranscriptional modification. miRNAs can suppress gene activity by binding to the mRNA of multiple target genes and preventing translation, cleaving the mRNA, or promoting mRNA degradation (Sato et al., 2011). An increasing number of miRNAs have been identified (e.g., miR-9, miR-137) which have specificity with regard to the range of target genes that are repressed by these molecules. Though perhaps not classically "epigenetic" in the sense that miRNAs typically target the product of transcription rather than interacting directly with the DNA, miRNAs play a significant role in gene regulatory networks and thus are an important consideration in the pathways linking genotype to phenotype.

D. Interaction between epigenetic pathways

Though a typical approach to the study of epigenetic mechanisms is to consider each separately, these factors are acting in concert to regulate gene expression and there are dynamic interactions between DNA methylation, posttranslational histone modifications, and miRNAs. For example, MBDs attract a protein complex to methylated DNA which includes HDACs (Fan and Hutnick, 2005; Razin, 1998). Consequently, DNA methylation often coincides with histone deacetylation. Pharmacological inhibition of HDACs has also been demonstrated to reduce levels of DNA methylation (Alonso-Aperte et al., 1999; Selker, 1998). miRNAs are known to be regulated by epigenetic modifications and to exert epigenetic modifications on the expression of HDACs and DNMTs (Sato et al., 2011). These interactions, combined with the complexity of each individual modification, make the prediction of gene expression based on any single epigenetic modification a challenging undertaking.

II. EPIGENETIC IMPACT OF SOCIAL EXPERIENCES

Though epigenetic mechanisms can confer a high degree of both plasticity and stability to gene expression patterns, the flexibility of these mechanisms (particularly DNA methylation) in response to environmental cues was initially thought to be limited to disruptions occurring during early embryonic development. However, it has been well established that stable individual differences in gene expression can be observed as a consequence of environmental experiences occurring across the lifespan (Covington et al., 2009; Lippmann et al., 2007; Meaney, 2001; Seckl and Meaney, 2004). Moreover, there is increasing support for the hypothesis that epigenetic mechanisms may be altered by a wide range of environmental events and thus mediate the long-term effects of the environment on biological and behavioral outcomes (Champagne, 2010; Jirtle and Skinner, 2007; Meaney and Szyf, 2005). In humans, the study of monozygotic twins, which has been a classic methodological tool within genetics, has provided intriguing insights into the potential for epigenetic plasticity in driving phenotypic divergence. DNA methylation and histone acetylation patterns are highly concordant among young twins (< 28 years of age) but diverge significantly among older twins (> 28 years of age), leading to the speculation that this divergence emerges across the lifespan in response to the unique environmental experiences of each twin (Fraga et al., 2005).

Though there are multiple features of the environment that can shape development, an intriguing question has been regarding the mechanisms through which our social experience come to be embedded in our biology. Decades of longitudinal and laboratory-based studies have highlighted the association between the quality of the social environment and neurobiological and behavioral outcomes (Ammerman et al., 1986; Miller et al., 2009; Pruessner et al., 2004; Sroufe, 2005; Trickett and McBride-Chang, 1995). The mediating role of epigenetic mechanisms in this association is becoming increasingly apparent, based primarily on evidence from animal models in which the quality of the social environment can be manipulated and brain region-specific epigenetic modifications assessed. In particular, these studies indicate epigenetic effects induced by the quality of postnatal mother–infant interactions and the experience of agonistic social encounters in adulthood.

A. Epigenetic effects of mother–infant interactions

The quality of social interactions occurring early in development can have a profound developmental effect. In humans, this effect is demonstrated by the severe cognitive and social deficits that emerge among infants that experience childhood neglect or abuse (Eluvathingal et al., 2006; MacLean, 2003; Trickett and McBride-Chang, 1995). Disruption to the parent–infant relationship can lead to impairments in attachment—the formation of a relationship with a primary caregiver which promotes social/emotional development—with subsequent effects on the risk of psychopathology (Bowlby, 1988; Egeland and Farber, 1984; Johnson et al., 2000). Parental sensitivity to infant cues is a key element within the formation of an attachment relationship, and variation in maternal sensitivity within the normal range has been found to predict stress sensitivity and negative affect among infants (Hane and Fox, 2006; Pederson et al., 1998). Through the use of neuroimaging techniques, the neurobiological consequences of disruption or variation in the mother–infant relationship are becoming increasingly apparent (Chugani et al., 2001; Eluvathingal et al., 2006; Pruessner et al., 2004). However, empirical studies of the molecular and epigenetic consequences of variation in the early social environment have been based primarily on rodent models of these experiences.

1. Variation in maternal care

Among most species, there are significant variations in the quality and frequency of parent–offspring interactions during the postnatal period (Champagne et al., 2003a, 2007; Fairbanks, 1989; Hane and Fox, 2006; Maestripieri, 1998). These individual differences in parental care can be influenced by environmental factors, such as stress, food availability, and the social context of the rearing environment (Champagne and Meaney, 2006, 2007; Curley et al., 2009; Gorman et al., 2002). Though there are species that engage in biparental or exclusive paternal care of offspring, for most species, the rearing of offspring is accomplished exclusively through mother–infant interactions. Among Long-Evans laboratory rats, the long-term and epigenetic consequences of maternal care have been explored and provide empirical support for the hypothesis that the quality of mother–infant interactions can lead to changes in DNA methylation and histone acetylation (Meaney and Szyf, 2005; Weaver et al., 2004). During the postnatal period, maternal care in rats is characterized by frequent nursing and licking/grooming (LG) of offspring. LG provides a source of tactile stimulation for offspring that alters physiology and promotes urination (Gubernick and Alberts, 1983; Sullivan et al., 1988a,b). Though a minimal level of LG is needed to achieve these physiological roles, it is evident that even within the stable conditions of laboratory housing, there are individual differences in the frequency with which lactating female rats will engage in this behavior (Champagne et al., 2003a). This variation can be used to study the impact of experiencing low levels of LG (Low LG) versus high levels of LG (High LG) on multiple neurobiological and behavioral outcomes in offspring (see Fig. 2.2). Male offspring of Low LG dams are observed to have a heightened hypothalamic–pituitary–adrenal (HPA) response to stress, impaired cognition, and reductions in hippocampal plasticity in adulthood (Caldji et al., 2000; Champagne et al., 2008; Liu et al., 1997, 2000). Female offspring reared by Low LG dams have decreased estrogen sensitivity in the medial preoptic area of the hypothalamus (MPOA) in adulthood and display reduced levels of care toward their own offspring (Champagne et al., 2001, 2003a,b). Moreover, cross-fostering studies indicate that these long-term outcomes are associated with the quality of maternal care experienced during the postnatal period rather than genetic or prenatal factors.

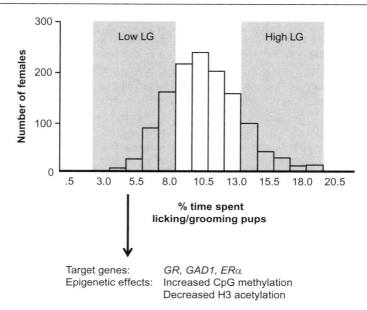

Figure 2.2. Summary of the epigenetic effects of natural variations in licking/grooming (LG) experienced during postnatal development. Lactating female rats engage in variation in the frequency of LG toward pups. Offspring reared by Low LG compared to High LG females can be compared on molecular, neurobiological, and behavioral measures. The experience of Low LG during infancy is associated with increased CpG methylation and decreased histone acetylation within the glucocorticoid receptor (GR) and glutamic acid decarboxylase (GAD1) gene promoter regions in the hippocampus of male offspring and the estrogen receptor alpha (ERα) promoter in the hypothalamus of female offspring.

To elucidate the potential role of epigenetic mechanisms in these maternal effects, gene regulation in the hippocampus of male offspring and MPOA of female offspring has been explored. The heightened HPA response to stress characteristic of offspring reared by Low LG dams is associated with impairments in HPA negative feedback (Liu et al., 1997). Within the hippocampus, glucocorticoid receptor (GR) levels are a key regulator of HPA negative feedback with a higher density of GR associated with enhanced ability to down-regulate the stress response (Sapolsky et al., 1985). Among adult male offspring of Low LG dams, there are reduced mRNA and protein levels of hippocampal GR which may account for the heightened stress responsivity observed in these offspring (Francis et al., 1999). Analysis of the GR 1_7 promoter region suggests that LG may exert these long-term effects on GR expression through DNA methylation (Weaver et al., 2004). Offspring that receive low levels of maternal

LG during the postnatal period have increased hippocampal GR 1_7 promoter DNA methylation at several CpG sites in this region. In particular, the binding site for the transcription factor NGFI-A (nerve growth factor-inducible protein A, also known as EGR-1 and ZIF268) is highly methylated among the offspring of Low LG dams and almost completely unmethylated among the offspring of High LG dams. These adult patterns of DNA methylation are not observed prenatally or at the time of birth but instead emerge during the first postnatal week, during which time the differential maternal LG behavior is most pronounced. Histone H3 acetylation at lysine 9 (H3K9) is also reduced within the GR 1_7 promoter of Low LG offspring as is the binding of NGFI-A to this region. These epigenetic effects can be reversed through ICV administration of the HDAC inhibitor trichostatin A into the hippocampal region of adult Low LG offspring (Weaver et al., 2004). Similarly, the ICV administration of methionine—a methyl donor—can increase GR 1_7 DNA methylation, decrease NGFI-A binding to the GR 1_7 promoter region, and decrease GR mRNA and protein levels in the adult offspring of High LG dams (Weaver et al., 2005)—illustrating the potential plasticity of these epigenetic effects.

Within the hippocampus of adult male offspring of Low LG dams, there are also elevated levels of DNMT1 mRNA (Zhang et al., 2010), providing a possible mechanism for global changes in DNA methylation and gene expression. Indeed, genome-wide expression assays indicate that over 900 genes are differentially expressed in the hippocampus as a function of maternal LG, and that overall there is increased gene expression among offspring of High LG compared to Low LG dams (Weaver et al., 2006). In addition to hippocampal GR, maternal LG has been found to alter expression of the enzyme glutamic acid decarboxylase (GAD1) gene, which may account for the maternal effects on hippocampal γ-aminobutyric acid circuits and receptor subunit composition (Caldji et al., 2003; Zhang et al., 2010). Analysis of DNA methylation within CpG sites of the GAD1 promoter region indicates that offspring reared by Low LG dams have heighted DNA methylation and reduced H3K9 acetylation in this region (Zhang et al., 2010). Thus, variation in maternal care can have widespread effects on gene expression which likely involve maternal effects on epigenetic modifications within the promoter regions of differentially expressed genes.

Maternal LG effects on female offspring have primarily focused on reproductive outcomes such as sexual and maternal behavior (Cameron et al., 2008a,b; Champagne et al., 2003a). Among female offspring of Low LG dams, there are decreased mRNA and protein levels of estrogen receptor alpha (ERα) in the MPOA, which may account for the reduced estrogen sensitivity and maternal care evident in these offspring (Champagne et al., 2001, 2003b). Analysis of levels of DNA methylation within the 1B promoter region of the ERα gene in MPOA tissue indicates that offspring of Low LG dams have elevated CpG methylation within this region (Champagne et al., 2006). As was the case for GR, these rearing environment-associated changes in DNA methylation alter the capacity for transcription factors to gain access to gene promoter regions. Among offspring of Low LG dams, there is decreased binding of the transcription factor STAT5a (signal transducer and activator of transcription 5A) to the ERα 1B promoter (Champagne et al., 2006). Thus, the differential DNA methylation associated with maternal LG likely has functional consequences for transcription factor-mediated upregulation of gene expression in the MPOA.

2. Maternal separation

The classic studies of Harlow demonstrated the dramatic effect of early life maternal deprivation on development in rhesus macaques (Harlow et al., 1965; Seay and Harlow, 1965), effects that are consistent with observations of the effects of social neglect on development in human infants. Epigenetic effects of maternal deprivation (through rearing of infant rhesus macaques in a nursery) have also been demonstrated. It has been noted in humans and rhesus that genetic polymorphisms in the serotonin transporter (5-HTT) gene promoter can lead to differential susceptibility to the effects of stress such that carriers of the short allele (which is predictive of less expression of 5-HTT) show heightened stress responsivity and increased risk of depression when exposed to increasing life stress, childhood maltreatment, or, in the case of rhesus, maternal deprivation (Aguilera et al., 2009; Caspi et al., 2003). DNA methylation levels within CpGs of the 5-HTT promoter in peripheral blood mononuclear cells collected from 90- to 120-day-old macaques were found to be higher in carriers of the short version of the 5-HTT allele (regardless of early rearing condition), and 5-HTT CpG methylation was found to interact with rearing condition to predict behavior. Elevated DNA methylation combined with maternal deprivation was found to predict heightened stress reactivity (Kinnally et al., 2010). Though it is yet unclear how to interpret the biological meaning of peripheral tissue measures of DNA methylation, these findings do suggest a potential role for epigenetic programming in the long-term effects of early life deficits in the social environment.

Maternal separation studies in mice have provided an opportunity to determine the brain region-specific epigenetic alterations that are induced through the experience of reductions in maternal care. In rodents, prolonged maternal separation (1–3 h per day) throughout the postnatal period has been found to induce changes in multiple neuroendocrine and neuropeptide systems and lead to heighted HPA response to stress (Lehmann and Feldon, 2000; Lippmann et al., 2007). Within the parvocellular neurons of the paraventricular hypothalamus (PVN), maternal separation has been found to induce persistent increases in arginine vasopressin (AVP) mRNA (Murgatroyd et al., 2009). Within the AVP gene, there are four regions rich in CpG islands that could potentially regulate gene expression through DNA methylation. Among male offspring that experienced maternal separation, DNA methylation is reduced at one of these four regions (CGI3) at 6 weeks, 3 months, and 1 year of age (Murgatroyd et al., 2009). This maternal effect may be PVN specific as rearing environment was not found to induce changes in AVP mRNA or DNA methylation in the supraoptic nucleus. Subsequent analyses indicated that the activation of MeCP2 (through phosphorylation of this MBD) may be a critical factor within these pathways, leading to AVP hypomethylation and increased AVP mRNA levels within the PVN. In a similar maternal separation paradigm in mice, male offspring exposed to prolonged separation from dams and reduced maternal care consequent to this manipulation were found to have increased DNA methylation within several CpG sites of the MeCP2 and cannabinoid receptor-1 (CB-1) genes and decreased CpG methylation within the corticotropin-releasing factor receptor-2 (CRFR2) gene (Franklin et al., 2010). These epigenetic effects were detected in sperm cells of maternally separated males. Interestingly, within the cortex of the female offspring of maternally separated males, increased DNA methylation in MeCP2 and CB-1 genes and decreased DNA methylation of CRFR2 were detected, possibly indicating the inheritance of epigenetic modifications by these offspring via the germ cells of maternally separated males.

3. Maternal abuse

The long-term consequences of childhood abuse that have been documented in humans and in primates illustrate the impact of agonistic social encounters occurring in infancy (Ammerman et al., 1986; Maestripieri, 2005). Laboratory rodent models of abuse have typically manipulated the quality of the rearing environment by restricting the amount of nesting materials available to dams. This disruption to the postnatal environment induces reductions in maternal care and an increased incidence of dams stepping on pups, aggressive grooming, and dragging of pups by a limb (Brunson et al., 2005; Ivy et al., 2008; Raineki et al., 2010). Similar to maternal separation, exposure to this rearing environment induces heightened stress responsivity and impairments in spatial memory (Avishai-Eliner et al., 2001; Brunson et al., 2005; Gilles et al., 1996; Raineki et al., 2010). Using a variation of this methodology with rats in which offspring are exposed to a brief encounter with an abusive nonbiological mother, the epigenetic effects of maternal abuse have been demonstrated. Male and female offspring that experience increased abusive behaviors in infancy have decreased brain derived neurotrophic factor (BDNF) mRNA levels within the prefrontal cortex in adulthood and within the IV promoter region of the BDNF gene, and abuse is associated with increased CpG methylation (Roth et al., 2009). Females that experience abusive rearing conditions are themselves more likely to engage in abusive behavior, and the offspring of abused females likewise have increased *BDNF* IV promoter DNA methylation in the prefrontal cortex, suggesting that environmentally induced epigenetic changes can be transmitted across generations.

4. Enhancing mother–infant interactions

Models of early life adversity are certainly most prevalent in the animal literature, yet there are also paradigms which lead to increased nurturing mother–infant interactions. Though prolonged maternal separation from offspring leads to reduced maternal care and heightened HPA response to stress in offspring, brief separations may stimulate maternal care, particularly LG, and attenuate offspring stress responsivity (Lehmann *et al.*, 2002; Meaney *et al.*, 1991). In rodents, brief maternal separation (also called *handling*) results in reductions in corticotropin-releasing factor (CRF) mRNA in the parvocellular neurons of the PVN that can be observed as earlier as postnatal day 9 (PN9) (Korosi *et al.*, 2010). Within the regulatory region of the rat *CRF* gene resides a binding element (NRSE) for the repressor neuron-restrictive silencer factor (NRSF) (Seth and Majzoub, 2001) and studies in human cell lines have shown that when NRSF is bound to this region, there is recruitment of cofactors and other enzymes/proteins involved in epigenetic regulation leading to the repression of gene expression (Zheng *et al.*, 2009). Among handled offspring, protein levels of NRSF are dramatically higher in PVN tissue at PN9 and throughout adulthood, suggesting a possible mechanism for handling induced reductions in *CRF* gene expression (Korosi *et al.*, 2010). Maternal care can likewise be stimulated through communal rearing, in which multiple lactating females rear offspring in a communal nest. In mice, this paradigm has been demonstrated to increase the frequency with which pups receive LG and nursing and leads to long-term changes in neurobiology and behavioral outcomes (Branchi, 2009; Curley *et al.*, 2009). Offspring that have been reared in a communal nest have increased hippocampal histone H3 acetylation at the *BDNF* I, IV, VI, and VII promoter regions, and this epigenetic modification may account for the increased *BDNF* levels observed among communally reared mice (Branchi *et al.*, 2011). Thus, enriching the early social environment through manipulations which increase maternal care can induce epigenetic modifications which alter neurodevelopmental outcomes.

5. Human studies of the epigenetic impact of childhood adversity

Though translating the findings of animal studies to better understand the impact of social experiences in humans is challenging, there is increasing evidence that epigenetic variation can be observed among individuals that have experienced childhood adversity. For example, analysis of postmortem human brain tissue suggests that individuals with a history of childhood abuse have decreased hippocampal GR mRNA associated with increased DNA methylation within the *GR* 1F promoter region when compared to nonabused subjects (McGowan *et al.*, 2009). In blood samples from orphans raised in institutions, genome-wide levels of DNA methylation have been found to be altered, with institution-raised children having overall higher levels of CpG methylation compared to children raised by their biological parents (Naumova *et al.*, 2011). The use of blood biomarkers in human studies of the epigenetic impact of social experiences may allow for the application of an epigenetic perspective to many critical questions regarding the timing, specificity, and stability of the effects of social environments on human development.

B. Adult social stress

Neurobiological and behavioral plasticity is not limited to infancy/early childhood and can certainly be observed in adulthood. In rodents, this plasticity can also be observed within epigenetic pathways. Though social interactions are important for normal development, and long-term social isolation can impair emotional behavior (Heidbreder *et al.*, 2000), the quality of those social encounters will be an important predictor of long-term health outcomes. In adult rodents, exposure to aggressive social interactions can have a significant impact on social behavior and induce depressive-like behaviors. The "social defeat" model has been used to induce these effects (Martinez *et al.*, 2002; Tamashiro *et al.*, 2005) and consists of placing an "intruder" into the cage of a "resident," typically a larger, dominant adult male (Miczek, 1979). In this paradigm, the intruder is exposed to repeated aggressive encounters and is defeated in these interactions. As is the case for early life adversity, this manipulation of the adult social environment results in reduced locomotion, social avoidance, and increased HPA activity (Blanchard *et al.*, 1993; Keeney and Hogg, 1999; Meerlo *et al.*, 1996; Raab *et al.*, 1986). Among socially defeated male mice, there are reduced hippocampal levels of *BDNF* that can be observed for a month following the experience of defeat (Tsankova *et al.*, 2006). Within the *BDNF* III and IV promoter regions, there is increased histone H3K27 dimethylation in socially defeated males which may account for the reduced *BDNF* expression. Histone deacetylase (*HDAC5*) mRNA levels are also found to be decreased in socially defeated males (Tsankova *et al.*, 2006). Differential levels of histone H3K27 dimethylation are also found across the genome within the nucleus accumbens (NAc), in response to both chronic social defeat and prolonged adult social isolation (Wilkinson *et al.*, 2009). Analysis of histone acetylation in the NAc indicates that H3K14 acetylation is initially decreased and then increased following chronic social defeat associated with decreases in *HDAC2* levels. Increased H3 acetylation in the hippocampus and infralimbic medial prefrontal cortex has also been observed in socially defeated rats (Hinwood et al., 2011; Hollis et al., 2010).

Individual differences in the effects of social defeat have been observed at both the behavioral and epigenetic level of analysis (see Fig. 2.3). Though social defeat has been found to induce long-term reductions in social approach behavior, there are some individuals that display resilience to this stressor (Wilkinson et al., 2009). Stress-susceptible mice have been found to have increased levels of CRF mRNA in the PVN and decreased CpG methylation within the CRF gene promoter, whereas stress-resilient mice were found to have no changes in CRF mRNA or DNA methylation of this gene (Elliott et al., 2010). In rats, individual differences in response to novelty predict differential epigenetic effects following exposure to social defeat. Among rats that are highly exploratory in a novel environment, social defeat results in decreased H3K14 acetylation, whereas among rats that engage in low levels of exploratory behavior, social defeat is associated with increased H3K14 acetylation (Hollis et al., 2011). Overall, these studies highlight the complex interactions between stress susceptibility, social experiences, and epigenetic pathways which will certainly be an important consideration in the study of epigenetic effects in humans.

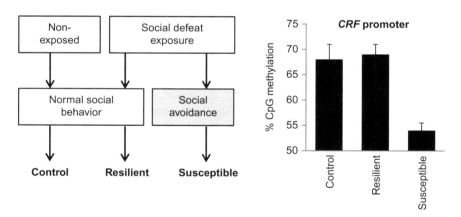

Figure 2.3. Summary of the effects of adult social defeat in mice on DNA methylation within the corticotropin-releasing factor (CRF) gene. The experience of social defeat results in lasting neurobiological and behavioral changes. In particular, this experience of social adversity typically leads to reduced social approach behavior (susceptible). However, some individuals do not exhibit impairments in social behavior following the experience of social defeat (resilient). Analysis of the degree of CpG methylation within the CRF gene promoter suggests that both nonexposed (control) and resilient mice have elevated levels of CRF CpG methylation, whereas susceptible mice have reduced CRF CpG methylation.

III. ROLE OF EPIGENETICS IN SHAPING SOCIAL BEHAVIOR

The previous section highlighted the growing evidence for the influence of social experiences on epigenetic mechanisms such as DNA methylation and histone modifications. In many cases, these epigenetic effects have consequences for social behavior. For example, the experience of Low LG or postnatal abuse leads to the emergence of these phenotypes in exposed individuals (Champagne, 2008; Roth et al., 2009). In the case of social defeat, epigenetic modifications may result in the sustained inhibition of social behavior among exposed individuals (Berton et al., 2006). Beyond these examples, there is increasing evidence for the role of epigenetic mechanisms in shaping social behavior derived from studies in which epigenetic pathways are directly manipulated or through modification of epigenetic pathways via nonsocial environmental cues which trigger sexual differentiation.

A. Epigenetic dysregulation and social behavior

The increased incidence of autism and other disorders characterized by social deficits has led to the development of diverse research approaches exploring both the genetic and epigenetic etiological pathways that account for disorders of the social brain. Genetic association studies have indicated a link between autism and mutations in the genes encoding the methyl-binding protein MeCP2 (Cukier et al., 2010; Schanen, 2006), the histone deacetylase HDAC4 (Williams et al., 2010), an H3K9 methyltransferase (EHMT1) (Balemans et al., 2010; Kleefstra et al., 2010), and the H3K4 demethylase JARID1C (Adegbola et al., 2008). These genetic effects have consequences for epigenetic chromatin remodeling in the autistic brain. Within neuronal cell populations in the prefrontal cortex, abnormalities in the pattern of H3K4 trimethylation have been detected in the postmortem brains of autistic individuals and these epigenetic modifications lead to altered mRNA levels of genes implicated in neurodevelopment (Shulha et al., 2011). Similarly, in mice, autism-like deficits in cognition and social behavior can be induced through targeted deletion of genes involved in epigenetic pathways. Mutation of the EHMT1 (histone-lysine N-methyltransferase) gene has been found to induce reductions in exploratory behavior and deficits in social play and suppress preference for social novelty in mice (Balemans et al., 2010), and in humans, mutation of this gene is associated with Kleefstra syndrome—a developmental disorder with autistic-like features (Kleefstra et al., 2010). In humans, mutations within the MeCP2 gene are associated with the development of Rett syndrome, characterized by cognitive decline in infancy and the development of stereotyped behaviors (Amir et al., 1999; Hagberg et al., 1983). Disruption to this gene in mice has indicated a role for MeCP2 in social behavior (Moretti and Zoghbi, 2006), and impairments in the functioning of *MeCP2* are associated with elevated H3 acetylation in the cortex (Shahbazian *et al.*, 2002). The importance of histone hyperacetylation in the etiology of social deficits has also been demonstrated using the HDAC inhibitor valproic acid. In mice, prenatal treatment with valproic acid has been found to induce cognitive impairments, increased anxiety-like behavior, and deficits in social interactions (Kataoka *et al.*, 2011). The hyperacetylation induced by valproic acid may lead to increased apoptosis in the developing prefrontal cortex with implications for cortical layering. Thus, epigenetic dysregulation, induced through genetic or pharmacologic factors, may have implications for a broad range of neurodevelopmental abnormalities.

B. Sexual differentiation

Hormonal signals occurring during development play a critical role in the sexual differentiation of the brain with implications for behavioral patterns exhibited by males and females. Across species, social play behavior has been found to be a sexually dimorphic behavior, with males typically displaying elevated levels of social play—particularly "rough-and-tumble" social interactions (Auger *et al.*, 2011; Meaney, 1989; Paukner and Suomi, 2008). Thus, sexual differentiation is a predictor of social behavior, and there is increasing evidence for the role of epigenetic mechanisms in the developmental process of this differentiation (Auger *et al.*, 2011; McCarthy *et al.*, 2009). The MPOA is a sexually dimorphic brain region that is typically larger in males than in females (Gorski *et al.*, 1978), and within this region, females have been found to have elevated levels of $ER\alpha$ mRNA and protein compared to males (Kurian *et al.*, 2010). The reduced transcription of $ER\alpha$ in males may be attributed to elevated levels of DNA methylation in the 1B promoter region of the $ER\alpha$ gene. However, if female rat pups are treated with estradiol on postnatal day 2, there is a decrease in $ER\alpha$ mRNA and this hormone-induced change in gene expression is associated with increased CpG methylation within the $ER\alpha$ gene promoter (Kurian *et al.*, 2010). Females can also become masculinized (decreased preoptic $ER\alpha$ mRNA, increased $ER\alpha$ 1B promoter methylation) by receiving licking-like tactile stimulation (using a paintbrush) during the postnatal period (Kurian *et al.*, 2010). Maternal LG has been observed to be directed more frequently to male pups during postnatal mother–infant interactions and may be an important behavioral pathway in the development of sexual differentiation (Moore, 1984; Moore and Morelli, 1979). Reductions in sexual dimorphism of neurobiological and behavioral measures has also been observed following exposure to endocrine disruptors, such as bisphenol A (BPA), and there is increasing evidence for the modification of epigenetic pathways by these compounds (Kundakovic and Champagne, 2011). BPA has been found to induce sex-specific effects on social behavior in juvenile mice, possibly through altered expression of the DNA methyltransferases

DNMT1 and *DNMT3a* (Wolstenholme *et al.*, 2011). Overall, these studies suggest that exploring the epigenetic basis of sexual differentiation and the processes (behavioral and hormonal) which alter differentiation may provide insights into the role of epigenetic mechanisms in shaping the social brain.

IV. CONCLUSIONS AND FUTURE DIRECTIONS

Epigenetic mechanisms play a critical role in development and may serve both to shape development in response to social experiences and to induce variation in social behavior. Thus, these molecular pathways illustrate the dynamic interactions between genes and environments that account for the origins of our unique neurodevelopmental and behavioral trajectories. The plasticity of DNA methylation and posttranslational histone modifications in response to both postnatal and adult social experiences suggests that these mechanisms may have evolved to allow organisms to adapt to changing environmental conditions. The persistence of behavioral, neurobiological, and epigenetic variation across generations that has been observed following manipulation of the social environment (Champagne, 2008; Franklin *et al.*, 2010; Roth *et al.*, 2009) may have implications for our notion of "inheritance" and the factors that contribute to the similarity between ancestors and descendants. Thus, our perspective on the factors driving both individual differences and the heritability of individual variation is increasingly combining both genetic and epigenetic pathways.

The field of behavioral epigenetics is rapidly expanding, and analysis of epigenetic mechanisms is being increasingly applied across species and taxa (Kucharski *et al.*, 2008; Meyer, 2011; Roy *et al.*, 2010). The incorporation of diverse species with unique life history characteristics will most certainly provide novel insights into both the convergent and divergent routes through which epigenetic plasticity drives phenotypic diversity. An important consideration within these studies will be the complex interactions between transcriptional, posttranscriptional, and posttranslational epigenetic modifications that are involved in the dynamic process of gene regulation. In addition, understanding the genetic basis of epigenetic variation, which is being increasingly explored in the context of social deficits in behavior, is a critical step in determining the evolutionary basis of epigenetic variation.

Acknowledgment

This research was supported by Grant Number DP2OD001674 from the Office of the Director, National Institutes of Health.

References

Adegbola, A., Gao, H., Sommer, S., and Browning, M. (2008). A novel mutation in JARID1C/SMCX in a patient with autism spectrum disorder (ASD). *Am. J. Med. Genet. A* **146A**, 505–511.

Aguilera, M., Arias, B., Wichers, M., Barrantes-Vidal, N., Moya, J., Villa, H., van Os, J., Ibáñez, M. I., Ruipérez, M. A., Ortet, G., and Fañanás, L. (2009). Early adversity and 5-HTT/BDNF genes: New evidence of gene? Environment interactions on depressive symptoms in a general population. *Psychol. Med.* **39**, 1425–1432.

Alonso-Aperte, E., Ubeda, N., Achon, M., Perez-Miguelsanz, J., and Varela-Moreiras, G. (1999). Impaired methionine synthesis and hypomethylation in rats exposed to valproate during gestation. *Neurology* **52**, 750–756.

Amir, R. E., Van den Veyver, I. B., Wan, M., Tran, C. Q., Francke, U., and Zoghbi, H. Y. (1999). Rett syndrome is caused by mutations in X-linked MECP2, encoding methyl-CpG-binding protein 2. *Nat. Genet.* **23**, 185–188.

Ammerman, R. T., Cassisi, J. E., Hersen, M., and van Hasselt, V. B. (1986). Consequences of physical abuse and neglect in children. *Clin. Psychol. Rev.* **6**, 291–310.

Auger, A. P., Jessen, H. M., and Edelmann, M. N. (2011). Epigenetic organization of brain sex differences and juvenile social play behavior. *Horm. Behav.* **59**, 358–363.

Avishai-Eliner, S., Gilles, E. E., Eghbal-Ahmadi, M., Bar-El, Y., and Baram, T. Z. (2001). Altered regulation of gene and protein expression of hypothalamic-pituitary-adrenal axis components in an immature rat model of chronic stress. *J. Neuroendocrinol.* **13**, 799–807.

Balemans, M. C., Huibers, M. M., Eikelenboom, N. W., Kuipers, A. J., van Summeren, R. C., Pijpers, M. M., Tachibana, M., Shinkai, Y., van Bokhoven, H., and Van der Zee, C. E. (2010). Reduced exploration, increased anxiety, and altered social behavior: Autistic-like features of euchromatin histone methyltransferase 1 heterozygous knockout mice. *Behav. Brain Res.* **208**, 47–55.

Barski, A., Cuddapah, S., Cui, K., Roh, T. Y., Schones, D. E., Wang, Z., Wei, G., Chepelev, I., and Zhao, K. (2007). High-resolution profiling of histone methylations in the human genome. *Cell* **129**, 823–837.

Berton, O., McClung, C. A., Dileone, R. J., Krishnan, V., Renthal, W., Russo, S. J., Graham, D., Tsankova, N. M., Bolanos, C. A., Rios, M., Monteggia, L. M., Self, D. W., et al. (2006). Essential role of BDNF in the mesolimbic dopamine pathway in social defeat stress. *Science* **311**, 864–868.

Blanchard, D. C., Sakai, R. R., McEwen, B., Weiss, S. M., and Blanchard, R. J. (1993). Subordination stress: Behavioral, brain, and neuroendocrine correlates. *Behav. Brain Res.* **58**, 113–121.

Bowlby, J. (1988). A Secure Base: Parent-Child Attachment and Healthy Human Development. Basic Books, New York.

Branchi, I. (2009). The mouse communal nest: Investigating the epigenetic influences of the early social environment on brain and behavior development. *Neurosci. Biobehav. Rev.* **33**, 551–559.

Branchi, I., Karpova, N. N., D'Andrea, I., Castren, E., and Alleva, E. (2011). Epigenetic modifications induced by early enrichment are associated with changes in timing of induction of BDNF expression. *Neurosci. Lett.* **495**, 168–172.

Brunson, K. L., Kramár, E., Lin, B., Chen, Y., Colgin, L. L., Yanagihara, T. K., Lynch, G., and Baram, T. Z. (2005). Mechanisms of late-onset cognitive decline after early-life stress. *J. Neurosci.* **25**, 9328–9338.

Caldji, C., Diorio, J., and Meaney, M. J. (2000). Variations in maternal care in infancy regulate the development of stress reactivity. *Biol. Psychiatry* **48**, 1164–1174.

Caldji, C., Diorio, J., and Meaney, M. J. (2003). Variations in maternal care alter GABA(A) receptor subunit expression in brain regions associated with fear. *Neuropsychopharmacology* **28**, 1950–1959.

Cameron, N., Del Corpo, A., Diorio, J., McAllister, K., Sharma, S., and Meaney, M. J. (2008a). Maternal programming of sexual behavior and hypothalamic-pituitary-gonadal function in the female rat. *PLoS One* **3**, e2210.

Cameron, N. M., Fish, E. W., and Meaney, M. J. (2008b). Maternal influences on the sexual behavior and reproductive success of the female rat. *Horm. Behav.* **54**, 178–184.

Caspi, A., Sugden, K., Moffitt, T. E., Taylor, A., Craig, I. W., Harrington, H., McClay, J., Mill, J., Martin, J., Braithwaite, A., and Poulton, R. (2003). Influence of life stress on depression: Moderation by a polymorphism in the 5-HTT gene. *Science* **301**, 386–389.

Champagne, F. A. (2008). Epigenetic mechanisms and the transgenerational effects of maternal care. *Front. Neuroendocrinol.* **29**, 386–397.

Champagne, F. A. (2010). Epigenetic influence of social experiences across the lifespan. *Dev. Psychobiol.* **52**, 299–311.

Champagne, F. A., and Meaney, M. J. (2006). Stress during gestation alters postpartum maternal care and the development of the offspring in a rodent model. *Biol. Psychiatry* **59**, 1227–1235.

Champagne, F. A., and Meaney, M. J. (2007). Transgenerational effects of social environment on variations in maternal care and behavioral response to novelty. *Behav. Neurosci.* **121**, 1353–1363.

Champagne, F., Diorio, J., Sharma, S., and Meaney, M. J. (2001). Naturally occurring variations in maternal behavior in the rat are associated with differences in estrogen-inducible central oxytocin receptors. *Proc. Natl. Acad. Sci. U.S.A.* **98**, 12736–12741.

Champagne, F. A., Francis, D. D., Mar, A., and Meaney, M. J. (2003a). Variations in maternal care in the rat as a mediating influence for the effects of environment on development. *Physiol. Behav.* **79**, 359–371.

Champagne, F. A., Weaver, I. C., Diorio, J., Sharma, S., and Meaney, M. J. (2003b). Natural variations in maternal care are associated with estrogen receptor alpha expression and estrogen sensitivity in the medial preoptic area. *Endocrinology* **144**, 4720–4724.

Champagne, F. A., Weaver, I. C., Diorio, J., Dymov, S., Szyf, M., and Meaney, M. J. (2006). Maternal care associated with methylation of the estrogen receptor-alpha1b promoter and estrogen receptor-alpha expression in the medial preoptic area of female offspring. *Endocrinology* **147**, 2909–2915.

Champagne, F. A., Curley, J. P., Keverne, E. B., and Bateson, P. P. (2007). Natural variations in postpartum maternal care in inbred and outbred mice. *Physiol. Behav.* **91**, 325–334.

Champagne, D. L., Bagot, R. C., van Hasselt, F., Ramakers, G., Meaney, M. J., de Kloet, E. R., Joels, M., and Krugers, H. (2008). Maternal care and hippocampal plasticity: Evidence for experience-dependent structural plasticity, altered synaptic functioning, and differential responsiveness to glucocorticoids and stress. *J. Neurosci.* **28**, 6037–6045.

Chugani, H. T., Behen, M. E., Muzik, O., Juhasz, C., Nagy, F., and Chugani, D. C. (2001). Local brain functional activity following early deprivation: A study of postinstitutionalized Romanian orphans. *Neuroimage* **14**, 1290–1301.

Covington, H. E., 3rd, Maze, I., LaPlant, Q. C., Vialou, V. F., Ohnishi, Y. N., Berton, O., Fass, D. M., Renthal, W., Rush, A. J., 3rd, Wu, E. Y., Ghose, S., Krishnan, V., et al. (2009). Antidepressant actions of histone deacetylase inhibitors. *J. Neurosci.* **29**, 11451–11460.

Cukier, H. N., Rabionet, R., Konidari, I., Rayner-Evans, M. Y., Baltos, M. L., Wright, H. H., Abramson, R. K., Martin, E. R., Cuccaro, M. L., Pericak-Vance, M. A., and Gilbert, J. R. (2010). Novel variants identified in methyl-CpG-binding domain genes in autistic individuals. *Neurogenetics* **11**, 291–303.

Curley, J. P., Davidson, S., Bateseon, P., and Champagne, F. A. (2009). Social enrichment during postnatal development induces transgenerational effects on emotional and reproductive behavior in mice. *Front. Behav. Neurosci.* **3**(25), 1–14.

Deaton, A. M., and Bird, A. (2011). CpG islands and the regulation of transcription. *Genes Dev.* **25**, 1010–1022.

Dokmanovic, M., Clarke, C., and Marks, P. A. (2007). Histone deacetylase inhibitors: Overview and perspectives. *Mol. Cancer Res.* **5,** 981–989.

Egeland, B., and Farber, E. A. (1984). Infant-mother attachment: Factors related to its development and changes over time. *Child Dev.* **55,** 753–771.

Elliott, E., Ezra-Nevo, G., Regev, L., Neufeld-Cohen, A., and Chen, A. (2010). Resilience to social stress coincides with functional DNA methylation of the Crf gene in adult mice. *Nat. Neurosci.* **13,** 1351–1353.

Eluvathingal, T. J., Chugani, H. T., Behen, M. E., Juhasz, C., Muzik, O., Maqbool, M., Chugani, D. C., and Makki, M. (2006). Abnormal brain connectivity in children after early severe socioemotional deprivation: A diffusion tensor imaging study. *Pediatrics* **117,** 2093–2100.

Fairbanks, L. A. (1989). Early experience and cross-generational continuity of mother-infant contact in vervet monkeys. *Dev. Psychobiol.* **22,** 669–681.

Fan, G., and Hutnick, L. (2005). Methyl-CpG binding proteins in the nervous system. *Cell Res.* **15,** 255–261.

Fraga, M. F., Ballestar, E., Paz, M. F., Ropero, S., Setien, F., Ballestar, M. L., Heine-Suner, D., Cigudosa, J. C., Urioste, M., Benitez, J., Boix-Chornet, M., Sanchez-Aguilera, A., *et al.* (2005). Epigenetic differences arise during the lifetime of monozygotic twins. *Proc. Natl. Acad. Sci. U.S. A.* **102,** 10604–10609.

Francis, D. D., Champagne, F. A., Liu, D., and Meaney, M. J. (1999). Maternal care, gene expression, and the development of individual differences in stress reactivity. *Ann. N. Y. Acad. Sci.* **896,** 66–84.

Franklin, T. B., Russig, H., Weiss, I. C., Graff, J., Linder, N., Michalon, A., Vizi, S., and Mansuy, I. M. (2010). Epigenetic transmission of the impact of early stress across generations. *Biol. Psychiatry* **68,** 408–415.

Gilles, E. E., Schultz, L., and Baram, T. Z. (1996). Abnormal corticosterone regulation in an immature rat model of continuous chronic stress. *Pediatr. Neurol.* **15,** 114–119.

Gorman, J. M., Mathew, S., and Coplan, J. (2002). Neurobiology of early life stress: Nonhuman primate models. *Semin. Clin. Neuropsychiatry* **7,** 96–103.

Gorski, R. A., Gordon, J. H., Shryne, J. E., and Southam, A. M. (1978). Evidence for a morphological sex difference within the medial preoptic area of the rat brain. *Brain Res.* **148,** 333–346.

Gubernick, D. J., and Alberts, J. R. (1983). Maternal licking of young: Resource exchange and proximate controls. *Physiol. Behav.* **31,** 593–601.

Hagberg, B., Aicardi, J., Dias, K., and Ramos, O. (1983). A progressive syndrome of autism, dementia, ataxia, and loss of purposeful hand use in girls: Rett's syndrome: Report of 35 cases. *Ann. Neurol.* **14,** 471–479.

Hane, A. A., and Fox, N. A. (2006). Ordinary variations in maternal caregiving influence human infants' stress reactivity. *Psychol. Sci.* **17,** 550–556.

Harlow, H. F., Dodsworth, R. O., and Harlow, M. K. (1965). Total social isolation in monkeys. *Proc. Natl. Acad. Sci. U.S.A.* **54,** 90–97.

Heidbreder, C. A., Weiss, I. C., Domeney, A. M., Pryce, C., Homberg, J., Hedou, G., Feldon, J., Moran, M. C., and Nelson, P. (2000). Behavioral, neurochemical and endocrinological characterization of the early social isolation syndrome. *Neuroscience* **100,** 749–768.

Hinwood, M., Tynan, R. J., Day, T. A., and Walker, F. R. (2011). Repeated social defeat selectively increases deltaFosB expression and histone H3 acetylation in the infralimbic medial prefrontal cortex. *Cereb. Cortex* **21,** 262–271.

Hollis, F., Wang, H., Dietz, D., Gunjan, A., and Kabbaj, M. (2010). The effects of repeated social defeat on long-term depressive-like behavior and short-term histone modifications in the hippocampus in male Sprague-Dawley rats. *Psychopharmacology (Berl.)* **211,** 69–77.

Hollis, F., Duclot, F., Gunjan, A., and Kabbaj, M. (2011). Individual differences in the effect of social defeat on anhedonia and histone acetylation in the rat hippocampus. *Horm. Behav.* **59,** 331–337.

Ivy, A. S., Brunson, K. L., Sandman, C., and Baram, T. Z. (2008). Dysfunctional nurturing behavior in rat dams with limited access to nesting material: A clinically relevant model for early-life stress. *Neuroscience* **154,** 1132–1142.

Jablonka, E., and Lamb, M. J. (2002). The changing concept of epigenetics. *Ann. N. Y. Acad. Sci.* **981,** 82–96.

Jenuwein, T., and Allis, C. D. (2001). Translating the histone code. *Science* **293,** 1074–1080.

Jirtle, R. L., and Skinner, M. K. (2007). Environmental epigenomics and disease susceptibility. *Nat. Rev. Genet.* **8,** 253–262.

Johnson, J. G., Smailes, E. M., Cohen, P., Brown, J., and Bernstein, D. P. (2000). Associations between four types of childhood neglect and personality disorder symptoms during adolescence and early adulthood: Findings of a community-based longitudinal study. *J. Pers. Disord.* **14,** 171–187.

Jones, P. A., and Taylor, S. M. (1980). Cellular differentiation, cytidine analogs and DNA methylation. *Cell* **20,** 85–93.

Kataoka, S., Takuma, K., Hara, Y., Maeda, Y., Ago, Y., and Matsuda, T. (2011). Autism-like behaviours with transient histone hyperacetylation in mice treated prenatally with valproic acid. *Int. J. Neuropsychopharmacol.* 1–13.

Keeney, A. J., and Hogg, S. (1999). Behavioural consequences of repeated social defeat in the mouse: Preliminary evaluation of a potential animal model of depression. *Behav. Pharmacol.* **10,** 753–764.

Kinnally, E. L., Capitanio, J. P., Leibel, R., Deng, L., LeDuc, C., Haghighi, F., and Mann, J. J. (2010). Epigenetic regulation of serotonin transporter expression and behavior in infant rhesus macaques. *Genes Brain Behav.* **9,** 575–582.

Kleefstra, T., Nillesen, W. M., and Yntema, H. G. (2010). Kleefstra syndrome. *In* "GeneReviews [Internet]" (R. A. Pagon, T. C. Bird, C. R. Dolan, and K. Stephens, eds.). University of Washington, Seattle, WA.

Koch, C. M., Andrews, R. M., Flicek, P., Dillon, S. C., Karaoz, U., Clelland, G. K., Wilcox, S., Beare, D. M., Fowler, J. C., Couttet, P., James, K. D., Lefebvre, G. C., *et al.* (2007). The landscape of histone modifications across 1% of the human genome in five human cell lines. *Genome Res.* **17,** 691–707.

Korosi, A., Shanabrough, M., McClelland, S., Liu, Z. W., Borok, E., Gao, X. B., Horvath, T. L., and Baram, T. Z. (2010). Early-life experience reduces excitation to stress-responsive hypothalamic neurons and reprograms the expression of corticotropin-releasing hormone. *J. Neurosci.* **30,** 703–713.

Kucharski, R., Maleszka, J., Foret, S., and Maleszka, R. (2008). Nutritional control of reproductive status in honeybees via DNA methylation. *Science* **319,** 1827–1830.

Kundakovic, M., and Champagne, F. A. (2011). Epigenetic perspective on the developmental effects of bisphenol A. *Brain Behav. Immun.* **25,** 1084–1093.

Kurian, J. R., Olesen, K. M., and Auger, A. P. (2010). Sex differences in epigenetic regulation of the estrogen receptor-alpha promoter within the developing preoptic area. *Endocrinology* **151,** 2297–2305.

Legube, G., and Trouche, D. (2003). Regulating histone acetyltransferases and deacetylases. *EMBO Rep.* **4,** 944–947.

Lehmann, J., and Feldon, J. (2000). Long-term biobehavioral effects of maternal separation in the rat: Consistent or confusing? *Rev. Neurosci.* **11,** 383–408.

Lehmann, J., Pryce, C. R., Jongen-Relo, A. L., Stohr, T., Pothuizen, H. H., and Feldon, J. (2002). Comparison of maternal separation and early handling in terms of their neurobehavioral effects in aged rats. *Neurobiol. Aging* **23,** 457–466.

Li, E., Bestor, T. H., and Jaenisch, R. (1992). Targeted mutation of the DNA methyltransferase gene results in embryonic lethality. *Cell* **69,** 915–926.

Lippmann, M., Bress, A., Nemeroff, C. B., Plotsky, P. M., and Monteggia, L. M. (2007). Long-term behavioural and molecular alterations associated with maternal separation in rats. *Eur. J. Neurosci.* **25,** 3091–3098.

Liu, D., Diorio, J., Tannenbaum, B., Caldji, C., Francis, D., Freedman, A., Sharma, S., Pearson, D., Plotsky, P. M., and Meaney, M. J. (1997). Maternal care, hippocampal glucocorticoid receptors, and hypothalamic-pituitary-adrenal responses to stress. *Science* **277,** 1659–1662.

Liu, D., Diorio, J., Day, J. C., Francis, D. D., and Meaney, M. J. (2000). Maternal care, hippocampal synaptogenesis and cognitive development in rats. *Nat. Neurosci.* **3,** 799–806.

MacLean, K. (2003). The impact of institutionalization on child development. *Dev. Psychopathol.* **15,** 853–884.

Maestripieri, D. (1998). Parenting styles of abusive mothers in group-living rhesus macaques. *Anim. Behav.* **55,** 1–11.

Maestripieri, D. (2005). Early experience affects the intergenerational transmission of infant abuse in rhesus monkeys. *Proc. Natl. Acad. Sci. U.S.A.* **102,** 9726–9729.

Martinez, M., Calvo-Torrent, A., and Herbert, J. (2002). Mapping brain response to social stress in rodents with c-fos expression: A review. *Stress* **5,** 3–13.

McCarthy, M. M., Auger, A. P., Bale, T. L., De Vries, G. J., Dunn, G. A., Forger, N. G., Murray, E. K., Nugent, B. M., Schwarz, J. M., and Wilson, M. E. (2009). The epigenetics of sex differences in the brain. *J. Neurosci.* **29,** 12815–12823.

McGowan, P. O., Sasaki, A., D'Alessio, A. C., Dymov, S., Labonte, B., Szyf, M., Turecki, G., and Meaney, M. J. (2009). Epigenetic regulation of the glucocorticoid receptor in human brain associates with childhood abuse. *Nat. Neurosci.* **12,** 342–348.

Meaney, M. J. (1989). The sexual differentiation of social play. *Psychiatr. Dev.* **7,** 247–261.

Meaney, M. J. (2001). Maternal care, gene expression, and the transmission of individual differences in stress reactivity across generations. *Annu. Rev. Neurosci.* **24,** 1161–1192.

Meaney, M. J., and Szyf, M. (2005). Environmental programming of stress responses through DNA methylation: Life at the interface between a dynamic environment and a fixed genome. *Dialogues Clin. Neurosci.* **7,** 103–123.

Meaney, M. J., Mitchell, J. B., Aitken, D. H., Bhatnagar, S., Bodnoff, S. R., Iny, L. J., and Sarrieau, A. (1991). The effects of neonatal handling on the development of the adrenocortical response to stress: Implications for neuropathology and cognitive deficits in later life. *Psychoneuroendocrinology* **16,** 85–103.

Meerlo, P., Overkamp, G. J., Benning, M. A., Koolhaas, J. M., and Van den Hoofdakker, R. H. (1996). Long-term changes in open field behaviour following a single social defeat in rats can be reversed by sleep deprivation. *Physiol. Behav.* **60,** 115–119.

Meyer, P. (2011). DNA methylation systems and targets in plants. *FEBS Lett.* **585,** 2008–2015.

Miczek, K. A. (1979). A new test for aggression in rats without aversive stimulation: Differential effects of d-amphetamine and cocaine. *Psychopharmacology (Berl.)* **60,** 253–259.

Miller, G. E., Chen, E., Fok, A. K., Walker, H., Lim, A., Nicholls, E. F., Cole, S., and Kobor, M. S. (2009). Low early-life social class leaves a biological residue manifested by decreased glucocorticoid and increased proinflammatory signaling. *Proc. Natl. Acad. Sci. U.S.A.* **106,** 14716–14721.

Monneret, C. (2005). Histone deacetylase inhibitors. *Eur. J. Med. Chem.* **40,** 1–13.

Moore, C. L. (1984). Maternal contributions to the development of masculine sexual behavior in laboratory rats. *Dev. Psychobiol.* **17,** 347–356.

Moore, C. L., and Morelli, G. A. (1979). Mother rats interact differently with male and female offspring. *J. Comp. Physiol. Psychol.* **93,** 677–684.

Moretti, P., and Zoghbi, H. Y. (2006). MeCP2 dysfunction in Rett syndrome and related disorders. *Curr. Opin. Genet. Dev.* **16,** 276–281.

Murgatroyd, C., Patchev, A. V., Wu, Y., Micale, V., Bockmuhl, Y., Fischer, D., Holsboer, F., Wotjak, C. T., Almeida, O. F., and Spengler, D. (2009). Dynamic DNA methylation programs persistent adverse effects of early-life stress. *Nat. Neurosci.* **12,** 1559–1566.

Nabel, C. S., and Kohli, R. M. (2011). Molecular biology. Demystifying DNA demethylation. *Science* **333,** 1229–1230.

Naumova, O. Y., Lee, M., Koposov, R., Szyf, M., Dozier, M., and Grigorenko, E. L. (2012). Differential patterns of whole-genome DNA methylation in institutionalized children and children raised by their biological parents. *Dev. Psychopathol.* **24**(1), 143–155.

Okano, M., Bell, D. W., Haber, D. A., and Li, E. (1999). DNA methyltransferases Dnmt3a and Dnmt3b are essential for de novo methylation and mammalian development. *Cell* **99,** 247–257.

Paukner, A., and Suomi, S. J. (2008). Sex differences in play behavior in juvenile tufted capuchin monkeys (Cebus apella). *Primates* **49,** 288–291.

Pederson, D. R., Gleason, K. E., Moran, G., and Bento, S. (1998). Maternal attachment representations, maternal sensitivity, and the infant-mother attachment relationship. *Dev. Psychol.* **34,** 925–933.

Peterson, C. L., and Laniel, M. A. (2004). Histones and histone modifications. *Curr. Biol.* **14,** R546–551.

Pruessner, J. C., Champagne, F., Meaney, M. J., and Dagher, A. (2004). Dopamine release in response to a psychological stress in humans and its relationship to early life maternal care: A positron emission tomography study using [11C]raclopride. *J. Neurosci.* **24,** 2825–2831.

Raab, A., Dantzer, R., Michaud, B., Mormede, P., Taghzouti, K., Simon, H., and Le Moal, M. (1986). Behavioural, physiological and immunological consequences of social status and aggression in chronically coexisting resident-intruder dyads of male rats. *Physiol. Behav.* **36,** 223–228.

Raineki, C., Moriceau, S., and Sullivan, R. M. (2010). Developing a neurobehavioral animal model of infant attachment to an abusive caregiver. *Biol. Psychiatry* **67,** 1137–1145.

Razin, A. (1998). CpG methylation, chromatin structure and gene silencing-a three-way connection. *EMBO J.* **17,** 4905–4908.

Roth, T. L., Lubin, F. D., Funk, A. J., and Sweatt, J. D. (2009). Lasting epigenetic influence of early-life adversity on the BDNF gene. *Biol. Psychiatry* **65,** 760–769.

Roy, S., Ernst, J., Kharchenko, P. V., Kheradpour, P., Negre, N., Eaton, M. L., Landolin, J. M., Bristow, C. A., Ma, L., Lin, M. F., Washietl, S., Arshinoff, B. I., et al. (2010). Identification of functional elements and regulatory circuits by Drosophila modENCODE. *Science* **330,** 1787–1797.

Santos, F., Hendrich, B., Reik, W., and Dean, W. (2002). Dynamic reprogramming of DNA methylation in the early mouse embryo. *Dev. Biol.* **241,** 172–182.

Sapolsky, R. M., Meaney, M. J., and McEwen, B. S. (1985). The development of the glucocorticoid receptor system in the rat limbic brain III. Negative-feedback regulation. *Brain Res.* **350,** 169–173.

Sato, F., Tsuchiya, S., Meltzer, S. J., and Shimizu, K. (2011). MicroRNAs and epigenetics. *FEBS J.* **278,** 1598–1609.

Schanen, N. C. (2006). Epigenetics of autism spectrum disorders. *Hum. Mol. Genet.* **15**(Spec. No. 2), R138–150.

Seay, B., and Harlow, H. F. (1965). Maternal separation in the rhesus monkey. *J. Nerv. Ment. Dis.* **140,** 434–441.

Seckl, J. R., and Meaney, M. J. (2004). Glucocorticoid programming. *Ann. N. Y. Acad. Sci.* **1032,** 63–84.

Selker, E. U. (1998). Trichostatin A causes selective loss of DNA methylation in Neurospora. *Proc. Natl. Acad. Sci. U.S.A.* **95,** 9430–9435.

Seth, K. A., and Majzoub, J. A. (2001). Repressor element silencing transcription factor/neuron-restrictive silencing factor (REST/NRSF) can act as an enhancer as well as a repressor of corticotropin-releasing hormone gene transcription. *J. Biol. Chem.* **276,** 13917–13923.

Shahbazian, M., Young, J., Yuva-Paylor, L., Spencer, C., Antalffy, B., Noebels, J., Armstrong, D., Paylor, R., and Zoghbi, H. (2002). Mice with truncated MeCP2 recapitulate many Rett syndrome features and display hyperacetylation of histone H3. *Neuron* **35**, 243–254.

Shulha, H. P., Cheung, I., Whittle, C., Wang, J., Virgil, D., Lin, C. L., Guo, Y., Lessard, A., Akbarian, S., and Weng, Z. (2011). Epigenetic signatures of autism: Trimethylated H3K4 landscapes in prefrontal neurons. *Arch. Gen. Psychiatry* **1236**, 30–43.

Sroufe, L. A. (2005). Attachment and development: A prospective, longitudinal study from birth to adulthood. *Attach. Hum. Dev.* **7**, 349–367.

Sullivan, R. M., Shokrai, N., and Leon, M. (1988a). Physical stimulation reduces the body temperature of infant rats. *Dev. Psychobiol.* **21**, 225–235.

Sullivan, R. M., Wilson, D. A., and Leon, M. (1988b). Physical stimulation reduces the brain temperature of infant rats. *Dev. Psychobiol.* **21**, 237–250.

Tamashiro, K. L., Nguyen, M. M., and Sakai, R. R. (2005). Social stress: From rodents to primates. *Front. Neuroendocrinol.* **26**, 27–40.

Trickett, P., and McBride-Chang, C. (1995). The developmental impact of different forms of child abuse and neglect. *Dev. Rev.* **15**, 11–37.

Tsankova, N. M., Berton, O., Renthal, W., Kumar, A., Neve, R. L., and Nestler, E. J. (2006). Sustained hippocampal chromatin regulation in a mouse model of depression and antidepressant action. *Nat. Neurosci.* **9**, 519–525.

Turek-Plewa, J., and Jagodzinski, P. P. (2005). The role of mammalian DNA methyltransferases in the regulation of gene expression. *Cell. Mol. Biol. Lett.* **10**, 631–647.

Turner, B. (2001). Chromatin and Gene Regulation. Blackwell Science Ltd, Oxford.

Waddington, C. H. (1942). The epigenotype. *Endeavour* **1**, 18–20.

Weaver, I. C., Cervoni, N., Champagne, F. A., D'Alessio, A. C., Sharma, S., Seckl, J. R., Dymov, S., Szyf, M., and Meaney, M. J. (2004). Epigenetic programming by maternal behavior. *Nat. Neurosci.* **7**, 847–854.

Weaver, I. C., Champagne, F. A., Brown, S. E., Dymov, S., Sharma, S., Meaney, M. J., and Szyf, M. (2005). Reversal of maternal programming of stress responses in adult offspring through methyl supplementation: Altering epigenetic marking later in life. *J. Neurosci.* **25**, 11045–11054.

Weaver, I. C., Meaney, M. J., and Szyf, M. (2006). Maternal care effects on the hippocampal transcriptome and anxiety-mediated behaviors in the offspring that are reversible in adulthood. *Proc. Natl. Acad. Sci. U.S.A.* **103**, 3480–3485.

Wilkinson, M. B., Xiao, G., Kumar, A., LaPlant, Q., Renthal, W., Sikder, D., Kodadek, T. J., and Nestler, E. J. (2009). Imipramine treatment and resiliency exhibit similar chromatin regulation in the mouse nucleus accumbens in depression models. *J. Neurosci.* **29**, 7820–7832.

Williams, S. R., Aldred, M. A., Der Kaloustian, V. M., Halal, F., Gowans, G., McLeod, D. R., Zondag, S., Toriello, H. V., Magenis, R. E., and Elsea, S. H. (2010). Haploinsufficiency of HDAC4 causes brachydactyly mental retardation syndrome, with brachydactyly type E, developmental delays, and behavioral problems. *Am. J. Hum. Genet.* **87**, 219–228.

Wolstenholme, J. T., Taylor, J. A., Shetty, S. R., Edwards, M., Connelly, J. J., and Rissman, E. F. (2011). Gestational exposure to low dose bisphenol A alters social behavior in juvenile mice. *PLoS One* **6**, e25448.

Wu, S. C., and Zhang, Y. (2010). Active DNA demethylation: Many roads lead to Rome. *Nat. Rev. Mol. Cell Biol.* **11**, 607–620.

Zhang, T. Y., Hellstrom, I. C., Bagot, R. C., Wen, X., Diorio, J., and Meaney, M. J. (2010). Maternal care and DNA methylation of a glutamic acid decarboxylase 1 promoter in rat hippocampus. *J. Neurosci.* **30**, 13130–13137.

Zheng, D., Zhao, K., and Mehler, M. F. (2009). Profiling RE1/REST-mediated histone modifications in the human genome. *Genome Biol.* **10**, R9.

3

One, Two, and Many—A Perspective on What Groups of *Drosophila melanogaster* Can Tell Us About Social Dynamics

Jonathan Schneider, Jade Atallah, and Joel D. Levine
Department of Biology, University of Toronto at Mississauga, Mississauga, Ontario, Canada

ABSTRACT

In the natural world, interactions between individuals occur in groups: an individual must recognize others, identify social opportunities, and discriminate among these options to engage in an interactive behavior. The presence of the group is known to exert an influence on individual group members, and this influence may feed back through the individual to affect behavior across the group. Such feedback has been observed in *Drosophila melanogaster*, for example, when mating frequency increases in groups composed of mixed strains compared to homogenous groups (Krupp *et al.*, 2008 and Billeter *et al.* 2012). A working hypothesis is that social processes—to recognize, identify, discriminate, and engage—are innate. They rely on a combination of genetic inheritance,

Advances in Genetics, Vol. 77
0065-2660/12 $35.00
http://dx.doi.org/10.1016/B978-0-12-387687-4.00003-9

molecular interactions, and cell circuitry that produce neural and immunological responses. Here, we discuss studies that emphasize social interactions in four categories in *Drosophila melanogaster*: learning, circadian clocks, aggression, and mating. We also speculate that a systems-level network approach to the study of *Drosophila* groups will be instrumental in understanding the genetic basis of emergent group-level behavior. © 2012, Elsevier Inc.

I. INTRODUCTION

> The problem of tracing the emergence of multidimensional behavior from the genes is a challenge that may not become obsolete so soon.
>
> Benzer (1971)

Although classified as a solitary insect species, *Drosophila melanogaster* displays a surprising array of behavioral interactions that are widely accepted as social: they court and mate; they fight over resources; they communicate using chemical, auditory, and tactile cues; they aggregate; they disperse; they synchronize their daily activity to one another; and there have been reports that they engage in observational learning to frame social expectations (Hall, 2002; Kravitz and Huber, 2003; Lefranc *et al.*, 2001; Levine *et al.*, 2002; Mery *et al.*, 2009; Sarin and Dukas, 2009; Suh *et al.*, 2004).

When a behavioral assay involves only two flies, such as with assays of courtship and copulation, there is no choice about interactive partners. The decision to interact, or not, depends on the qualities and state of both interactee and interactor. The interaction itself is determined by the two individuals, and any behaviors exhibited by the dyad are modulations of their interaction. However, once in a group, the presence of many individuals produces a complex social context, and it becomes necessary for an interactor to connect with one among many potential interactees. Understanding patterns of copulation or any social interaction within the group may involve more than scaling up from one or two to many because the other members of the group form an environment that can modulate, facilitate, or disrupt the dyad (e.g., competition for a mate can preclude an individual's copulatory behavior).

In addition to affecting patterns of behavior, the group also influences its individual members directly. This group influence may affect gene transcription, neuronal morphology, pheromonal communication, and subsequent behavior (Billeter *et al.*, 2009; Billeter *et al.* 2012; Donlea *et al.*, 2009; Eddison *et al.*, 2011; Kent *et al.*, 2008; Krupp *et al.*, 2008): as the group members change behavioral patterns and pheromonal profiles, the character of the social milieu can change (no detailed study of dynamics, but see Kent *et al.*, 2008). In this way, the variety of behaviors that are known to be influenced by, and exert an influence on, the social context create complex systems with many interactions and feedback loops between individuals. The dynamics of such complex systems not only modulate an individual's behavior and gene expression (see indirect genetic effects; Moore *et al.*, 1997), but such dynamics may also lead to emergent, group-level behavioral phenotypes.

The challenge of investigating group dynamics is therefore one of dissecting the many effects that co-occur within, and potentially modulate, the group. At a minimum, three social factors exert effects on behaviors within the group: number of potential interactors, genotypic composition, and timing of social experience. In a homogeneous setting composed of individuals from the same strain or genotype, manipulating the number of individuals in the group can reveal nonlinear behavioral response curves and threshold values for social, group-level effects (Saltz and Foley, 2011; Wang and Anderson, 2010). Mixing together groups can uncover emergent dynamics that are not seen in the respective homogeneous settings and may uncover group-level responses to strain or species recognition and competition (Kent *et al.*, 2008; Krupp *et al.*, 2008; Levine *et al.*, 2002). The influence of the timing of the social context, whether prior to or during observation, can suggest how this change in the social surrounding is being integrated, via developmental changes, learning, or circadian clocks, for example.

The group can therefore be seen as a complex system that may give rise to the emergence of group-level behavioral patterns. While the *cues* underlying group behaviors continue to be discovered (such as sex and species "tags" (Billeter *et al.*, 2009)), the *mechanisms* of these complex systems—the organization, how the flies assemble and disseminate information—are still a black box. If the rules of group organization are coded genetically, the difficulty in dissecting the path from gene transcription through behavior to group organization prompts the question asked by Michael Bate in another context, "[. . .] how [are] we going to look for the genes that regulate [social] behavior?" (Bate, 1998). Bate (1998) was asking about the genetic substrate required for a maggot to develop the ability to crawl; here, we are asking about the genetic requirements for participation in complex social groups.

In the first part of this perspective, we consider several broad behavioral categories and emphasize effects elicited by social context. We also present several cases of groups exhibiting emergent properties arising through the dynamic interactions of group members. While the investigation of such group-level phenotypes is not straightforward, their identification is becoming more and more frequent and can be done without specialized techniques.

In the second part, we advocate treating groups of flies as complex systems, with many interactions and feedback loops between individuals. These relationships may explain the emergence of group-level behavioral patterns and allow us to understand the mechanisms that generate them. We have begun to approach the biology of the fly at a group level using system-level tools from network theory. We speculate that such a systems-level approach to the study of *Drosophila* will be instrumental in answering our version of Bate's dilemma.

A. Caveat

We have elected to write this as a perspective piece. We focus on *D. melanogaster* with the qualification that the analysis of group dynamics is warranted and carried out in other animals as well. There are many important studies that will not be presented here. We hope the authors of those studies—many of whom are friends—will not feel slighted.

II. THE BEHAVIORAL EFFECTS OF SOCIAL CONTEXT

A. Social learning and memory

Whether an experiment relies on prior experience such as previous social contexts, competitive outcomes in aggression, or matings, the possibility of a social manipulation becoming a "conditioning" stimulus must be considered (Mery *et al.*, 2009; Sarin and Dukas, 2009). Although untested, molecular and cellular mechanisms of learning and memory could prove informative when investigating the pathways underlying social interactions (Ganguly-Fitzgerald *et al.*, 2006). Only a few studies have assessed the role of known learning substrates on social experience, yet the ability to learn and modify behavior as a consequence is likely to be a relevant feature of social experience. It is possible that learning pathways contribute to mechanisms of social engagement.

D. *melanogaster* has contributed insights into the genetics and neurobiology of learning. Many of these contributions were developed using olfactory learning assays such as the T-maze which rely on an index of learning and memory measured in groups (upward of 60 flies), but this feature of learning assays has not been a focus of study (Davis, 2004; Kahsai and Zars, 2011; Keene and Waddell, 2007; McGuire *et al.*, 2005). It is usually assumed—and why not?—that mechanisms of learning, especially learning of an operant task, apply to the individual. Here, we restrict our discussion to studies that directly evaluate the influence of the social environment on learning and memory of the individual.

Several studies have reported that individual flies display the ability to change their behavior by observing other flies within a group, a form of social learning. For instance, whereas a female will remain in closer proximity to a healthy male in comparison to a poorly fed male, she will change her preference if she detects another female's proximity to the unhealthy male (Mery *et al.*, 2009). This demonstrates that females acquire information by observing others and display selective responsiveness to this input. The ability to acquire such information from the social environment is also evident when females display mating preferences for males that are dusted with the same color as males previously observed copulating (Mery et al., 2009). This suggests that individuals are able to categorize socially acquired stimuli and remember them for subsequent mating choices. Selecting an appropriate egg-laying site is yet another process in which females exploit information provided by conspecifics (Sarin and Dukas, 2009). Individual females prefer food types as an egg-laying medium after they have experienced the food with mated but not virgin females. Moreover, a location preference for oviposition can be communicated without demonstrating egg laying on the medium itself (Battesti et al., 2012), indicating that naive individuals can recognize and learn from females who have been conditioned to prefer a particular egg-laying substrate. These studies demonstrate learned behavior based on the selective transfer of information between the individuals.

Despite such evidence supporting the idea that acquiring information from the social environment influences an individual's behavior in a group, the possibility that classical learning pathways explain this influence has not been well tested. One study has examined the influence of conspecifics using a classical olfactory learning and memory retrieval paradigm (Chabaud et al., 2009). In this assay, flies are trained to avoid an odor using shock and then subsequently evaluated on the avoidance of that odor to assess anesthesia-resistant memory (ARM) formation and retrieval. Flies trained in groups or in isolation were tested individually, and no differences were observed. However, subsequent testing showed that flies tested in groups achieve higher scores than flies tested individually on the ARM task. This improved performance does not involve nonspecific aggregation or herding because it is observed even when the accompanying conspecifics have been trained on different odors. In other words, the simple presence of conspecifics during ARM retrieval seems to improve performance. The aspect of the social environment that facilitates this memory retrieval is still unknown and deserves further research. Yet, these findings clearly demonstrate that the social environment influences ARM-dependent performance.

More generally, these studies raise questions such as: When groups of flies are trained how many of the flies are conditioned by the experimental stimulus, how many are following the lead of others, how many acquire conditioning via social interactions within the assay, and how much of their performance depends on a social contribution to retrieval? When group conditioning is tested, is it a sum of individual information? If information is distributed throughout the group, is there positive feedback between individual and group information? Data suggest that when individual appraisals contradict public information, individuals do not always ignore what they have learned (Chabaud et al., 2009; Mery et al., 2009). On the other hand, agreement with others in the group is not necessary for the socially mediated performance improvement in conditioning tasks (Chabaud *et al.*, 2009). This complex interplay between a fly's individual experience, social communication of information, a group's conditioned status, and the influence of the group on the individual are all potentially at play when groups are investigated. The propagation of conditioning (Battesti *et al.*, 2012) and facilitation of retrieval (Chabaud *et al.*, 2009) could play a role in emergent group-level phenotypes (see Billeter *et al.* (2012); Higgins *et al.* (2005); Schneider *et al.* (2012)).

B. Social synchronizing of activity and rest

Flies aggregate at dawn and dusk in the wild (Shorrocks, 1972; Wertheim *et al.*, 2005). The temporal organization of aggregation sets the stage for other social interactions, and it is based on circadian clock mechanisms in the adult brain. These clocks are thought to gate sensory excitability and enable an individual to anticipate key events in the environment. Individuals must be active, at the same time, in order to aggregate around food and to reproduce. Similarly, they must time their rest to ensure participation in group functions. Such biological synchrony can be coordinated by environmental cues such as the photoperiod, and the influence of the physical environment on the periodicity of the biological clock is well studied (Allada and Chung, 2011). Social synchrony might also be facilitated by the influence of social cues, but understanding the influence of social cues on the biological clock is just beginning.

Several studies have established that social groups influence the timing of various behaviors in flies, from locomotor activity bouts to circadian patterns of mating (Billeter *et al.*, 2012; Fujii *et al.*, 2007; Krupp *et al.*, 2008; Levine *et al.*, 2002; Lone and Sharma, 2011; Tauber *et al.*, 2003). Rhythmic patterns of behavior are coordinated within a group; flies maintained in groups are more synchronous in their individual patterns of activity than isolated flies under the same light:dark regimes (Levine *et al.*, 2002). This modulation of individual activity depends on the rhythmicity of flies in the groups and the ratio with which they are mixed. Arrhythmic mutants decrease synchrony among wild-type flies, and the strongest effects are seen on a rhythmic majority housed with an arrhythmic minority (Levine *et al.*, 2002; Lone and Sharma, 2011). These effects are mediated by unidentified olfactory cues (Levine *et al.*, 2002; Lone and Sharma, 2011). Temporal patterns of reproductive interactions are modified between cohabiting males and females (Fujii *et al.*, 2007). Same sex couples remain synchronous to isolated individuals, but male–female couples reset the timing of activity when housed together (Fujii *et al.*, 2007). Further, when groups of six males are maintained with six females, the temporal pattern of mating is regulated by the strain of the female: the female's pattern is evident whether placed with males of the same or different strains (Billeter *et al.*, 2012).

Although these studies suggest that flies may reset the clocks of other flies within a group, they do not prove that the clock mechanism is affected by the social environment. However, a subsequent study has shown that when groups are mixed in a way that is similar to the manipulation performed by Levine *et al.* (2002), the expression levels and amplitude of circadian clock genes in the head and in peripheral tissue respond to social manipulation (Krupp *et al.*, 2008). These findings demonstrate that an individual's circadian clock mechanism can be influenced by the social environment and suggest that synchrony is reflected in the transcriptional fluctuations that keep time. As noted, effects of these social manipulations on activity patterns depend on olfactory cues (Levine *et al.*, 2002; Lone and Sharma, 2011) and require intact rhythmic sensitivity of peripheral clock cells in the affected group members (Krupp *et al.*, 2008; Levine *et al.*, 2002).

A similar experimental approach demonstrated that social experience modifies individual sleep architecture, where the size of the group in which individuals are kept regulates the length of sleep bouts during the day (Ganguly-Fitzgerald *et al.*, 2006). This plasticity in sleep requires intact visual and olfactory communication between the group members. In these studies, a variety of genetic, molecular, and neural components have been associated with the group effect. Whole brain dopamine levels as well as the expression of a subset of short- and long-term memory genes correlate with this social plasticity in sleep (Ganguly-Fitzgerald *et al.*, 2006). Also, the expression of *rutabaga, period,* and *blistered* in a subset of circadian clock neurons is required for the display of experience-dependent variation in sleep architecture (Donlea *et al.*, 2009). Interestingly, the morphology of these clock neurons is also altered: an increase in the number of synaptic terminals correlates with the social experience that influences sleep (Donlea *et al.*, 2009). Thus, prior social experience influences circadian patterns of activity (Levine *et al.*, 2002; Lone and Sharma, 2011) and sleep (Ganguly-Fitzgerald *et al.*, 2006), neuronal morphology of clock cells in the brain (Donlea *et al.*, 2009) as well as patterns of gene expression associated with timekeeping (Krupp *et al.*, 2008). Taken together, these studies indicate that biological clock mechanisms not only regulate circadian behavior but are also modified by social stimuli.

These effects of the social group are not restricted to timing. Individual activity levels remain continuously susceptible to a variety of group properties such as genotype and gender (Higgins *et al.*, 2005). In one study, a complex metric was formed from measures of many behaviors such as walking, grooming, feeding, mating, courting, and fighting (Higgins *et al.*, 2005). This metric was used to determine the repeatability of intraindividual variation in isogenic groups using a repeated measures design. This study demonstrated that emergent characteristics of activity at the group-level explained as much variation in overall activity as the genotype of the group. However, the interpretation of this study is not straightforward because the "activity" metric itself is complicated. This metric does not discriminate between locomotion, sleep, and mating like the aforementioned studies. Instead, it reflects interplay between many or all behaviors and individuals of the group. It is tempting to ascribe such an emergent property as arising from the excitation and positive feedback of individuals interacting with one another. This study by Higgins *et al.* (2005) is especially noteworthy because the group-level activity is not easily predicted from the behavior of the individual members or from their genotype but from the ongoing interactions throughout the group.

Social groups can therefore influence patterns of behavioral activity by modulating the temporal structure of group interactions on a daily scale and within a given short-term sequence of events (Higgins *et al.*, 2005). Notably, this organization of behavior influenced by the group provides a glimpse of an emergent property based on complex patterns of behavioral activity. Further, variable patterns of behavioral activity are evident even when the genetic substrate has been isogenized, suggesting that such emergent properties have a stochastic component that operates separately from the genetic component.

C. Aggression

When flies meet, they may engage in a variety of offensive and defensive physical interactions, and highly detailed ethograms of these exchanges have been generated (Chen *et al.*, 2002). Aggression in flies was thought to be a competition for resources, and original studies on aggressive behavior were performed in a group setting (Dow and von Schilcher, 1975; Hoffmann, 1987a,b, 1988). Aggressive displays between pairs of males in groups were described as contests with a clear winner and loser. The underlying interpretation of early studies on fly aggression suggested the presence of a hierarchical group structure (Dow and von Schilcher, 1975). The benefit of aggressive contests was thought to be establishing a territorial priority over food patches in the presence of other males for access to females (Hoffmann, 1987a).

There is considerable phenotypic variation in this behavior, some of which is the result of genetic variation (Hoffmann, 1988), body size (Hoffmann, 1987b), and age (Hoffmann, 1990). Given the inherent social aspect of aggression (needing at least one other fly to aggress), it is perhaps not surprising that prior social experience also plays a prominent role in aggressive behavior. Rearing individuals in the presence of males and not females significantly reduces aggressive behavior (Hoffmann, 1990). This adjustment is not simply the "loser effect" (experienced losers mostly lose second fights (Yurkovic *et al.*, 2006)) of previous competitors, as individuals' aggression toward males with which they have been previously housed does not significantly differ from that displayed toward unfamiliar males (Hoffmann, 1990).

Subsequent studies of aggression have undergone a shift away from observing aggression within a group to studying aggression between isolated pairs of males. Consequently, efforts were allocated to establishing a simplified yet optimized paradigm in which a pair of males would fight. This allowed rapid advancement in the identification of the genetics and neurobiology (see Chan and Kravitz, 2007; Dierick, 2008; Dierick and Greenspan, 2006, 2007; Miczek *et al.*, 2007; Robin *et al.*, 2007).

Despite the great strides made in studying aggression, the current paradigm shift away from the group has hindered the analysis of aggression's influence on social organization (specifically the hierarchy described by Dow and von Schilcher, 1975). Unfortunately, studies investigating the "loser effect" are limited to a two-fight paradigm. This raises several questions. Is the "loser effect" cumulative? Can it be mitigated/erased by a victory? If the decrease in aggression and probability of winning is proportional to the amount of loses, a hierarchy seems inevitable. Even if winning mitigates the "loser effect," since flies appear to have individual as well as general history-dependent aggression, a loser–winner may still reduce the aggression toward his initial opponent, again creating a hierarchy, although one in which the dynamics of contest outcomes becomes pivotal. However, to address such hypotheses, the current aggression paradigm must expand past the two-fight stage, to three bouts (at a minimum) or a group (optimal).

Current studies are beginning to bridge the gap between dyadic aggressive contests and group-influenced aggression. One aspect of group-influenced aggression, the olfactory contributions of a group, has been shown to stimulate male–male aggression (Wang and Anderson, 2010). Wang and Anderson (2010) posit that *cis*-Vaccyl-Acetate (cVA) cue may play a dual role: not only is it a well-studied aggregation pheromone (Wertheim *et al.*, 2005), but it may be also acting as a dispersal cue when groups reach critical mass through aggression and territoriality. Wang and Anderson's (2010) model is parsimonious with regard to the independent finding of highly aggressive strains forming less dense groups (Saltz and Foley, 2011) and predicts that group density and group-level aggressiveness are part of a negative feedback mechanism linking group structure to individual behavior. This suggests that aggression may reduce the amount of cVA required to cause dispersal. However, cVA performs many functions, some inconsistent with others, a noteworthy topic beyond the scope of this perspective.

A full model of group-level aggression may therefore shed light on group composition. However, any such model must incorporate both the dynamic effect of aggression on density and the experience-dependent modulation of individual aggressiveness. Flies learn from and remember previous encounters (Yurkovic *et al.*, 2006), and the aggression of an individual at a given point in time is influenced by a complicated result of not only its previous conflicts but also the opponent's previous aggressive bouts (the results of which themselves are influenced by past opponents). While the depth of history that affects the current aggressiveness of a fly is still unclear, it is tempting to visualize the propagation of "wins" and "losses" as a network of aggressive bouts. The size of the group, the pattern, and outcome of these bouts affect not only an individual's propensity to aggress but also the density of the surrounding group.

This dynamic balance between winners and losers and the modulation of group density is complex. Quantifying the properties of such an aggressive network would allow the prediction of whether group-level aggression approaches a genetically determined equilibrium, or if a stochastic process takes shape (the former similar to examples below (Billeter *et al.*, 2012; Schneider *et al.*, 2012) and the latter similar to the group-level activity reported by Higgins *et al.*, 2005).

D. Mating, paternity, and offspring

We have discussed the social influences on learning, on synchronizing patterns of activity, and on aggression. These socially modulated behaviors also seem to contribute to reproductive decisions in *Drosophila*. Although courtship effort can also be considered as a component of fitness, we restrict our discussion to the simpler and more fitness-related measurements of copulation and paternity. Prior social experience has significant input on female mate choice (Mery *et al.*, 2009). Females prefer males of a specific color after observing males of that color copulate, and males colored differently get rejected. We also note a circadian aspect of mating, with wild-type flies exhibiting a cyclical mating rhythm under the control of the same clock genes underlying locomotor activity rhythms (Sakai *et al.*, 2002). Yet, even when in the same aggregate at the same time, a male must obtain a female despite conspecific male competitors. This competition is thought to involve aggression which has been tied to copulation, as the presence of a female increases aggressiveness, and winning is correlated to success in mating through food territoriality (Dow and von Schilcher, 1975; Hoffmann, 1987b).

Earlier studies point to evidence that females are also able to distinguish between the minority and the majority in heterogeneous groups of males through olfactory cues; these early studies dubbed this a "rare male effect" as preference was displayed toward the minority males (Ehrman, 1970, 1972). Recently, this idea of heterogeneous group effects on mating behavior was revisited. Within a 24-h period, the males in a genotypically homogeneous group mate fewer times than the males in a heterogeneous group (Billeter *et al.*, 2012; Krupp *et al.*, 2008). There is also a parallel influence of group composition on individual physiology (Kent *et al.*, 2008; Krupp *et al.*, 2008) and on the expression of clock genes as well as on the expression of genes involved in pheromone synthesis (Krupp *et al.*, 2008). These changes in gene expression are accompanied by changes in the production of key hydrocarbons, which in turn are known to play a major role in communication and reproductive behavior (Billeter *et al.*, 2009). The reproductive consequences associated with changes in hydrocarbons are further supported by studies where male group heterogeneity does not increase male matings when housed with olfactory mutant females and also indicates that this social effect is at least partly female driven (Billeter *et al.*, 2012).

A similar effect is observed when females of one strain are placed with a genotypically heterogeneous mix of males at varying ratios (Billeter *et al.*, 2012). Male mating success and female fecundity are affected by the interaction of male strain and group composition, and both are dependent on female strain. Surprisingly, when examining all offspring produced by the group, each male strain sires a consistent percentage of all offspring regardless of the ratio of heterogeneity in the group. This is particularly interesting given that this phenotype arises as a group property and cannot be detected in any one female within the group. This illustrates that while individual behavior can be variably responsive to the social environment, it can do so in a controlled fashion that can generate a consistent group phenotype. The mechanistic underpinnings of such regulation are still unknown. However, one possibility lies in the manipulation of last male sperm precedence as any individual female can potentially negate the effect of last male sperm precedence when copulating with multiple individuals (Singh *et al.*, 2002). Given unrestricted access to males in a group, a dilution effect is seen concerning the paternity patterns, and it is possible that females could modify the paternity according to the order, and the amount of copulations (Billeter *et al.*, 2012). While this possibly describes *how* a female fly modifies her offspring ratio, it does not address the mechanisms of the evaluation of and response to the social context. Are the pathways involved in evaluating the social context restricted to olfaction? Is the female quantifying the masculine environment? What pathways are involved in the female's subsequent predisposition to mating with a certain male? Does the male exhibit any influence through seminal fluid transfer? Are females observing conspecific matings to learn the patterns of copulation of the group? It is also noteworthy that this is only observed in one strain, while the other strain studied simply maintains an elevated frequency of mating in any social environment. This elevation does not produce a fixed ratio of offspring but may increase offspring genetic variability by indiscriminately increasing the number of matings. This strain-dependent difference in female reproductive strategies may eventually shed light on the genetic complications associated with the rare male effect, which is more prominent in certain strains than others, and may highlight shared mechanisms underlying strain-specific reproductive decisions.

Given the emergent property of offspring genetic diversity, it is unlikely that these reproductive decisions are based solely on an individual's chemical sampling of the heterogeneity of the group. Rather, we hypothesize that the process is a dynamic one, determined not only by a female's copulatory history but also by the prior and ongoing copulations within the social context. This process may be affecting both males and females and may be partly explained by seminal fluid transfer: not only is any male being modulated by the presence of potential con- and hetero-specific rivals (Bretman *et al.*, 2009), but the female's physiology/behavior could also be modified, which would in turn affect future copulations/sperm transfer (Lupold *et al.*, 2011). In tandem, a female's observation of her conspecific's choice in mating partners may affect her predisposition to mate with a given type of male. Investigations into these potential processes require an approach that incorporates the current, as well as the past, patterns of copulation as this process takes shape. These complex histories of prior mating experience, visual conspecific learning, physiological changes, all potentially play a role, and deciphering the patterns of the group will be pivotal in investigating the group-level mating phenotype and its consequences.

Although the emphasis on group dynamics is relatively new, several themes are already emerging. For example, the social environment affects gene expression and metabolism within individuals. Additionally, interactive behaviors, such as the frequency or temporal distribution of mating, vary when the group is manipulated. The implications of such variability in an interactive behavior may not be trivial. Moreover, the distinction between an individual and its environment may be a useful experimental device, but it too could be artificially limiting. We have been developing a method to examine the group as a social interaction network, a complex group-level entity. Our results suggest the possibility that traditional social behaviors—like mating and fighting—are influenced by the details of an underlying social interaction network.

III. SYSTEM-LEVEL AND NETWORK APPROACHES TO SOCIAL CONTEXT

The emergent group-level phenotypes highlighted above suggest that groups of flies undergo dynamic organization. Such complex systems display qualities that are not present at the individual level but are the additive and nonadditive effects of individual interactions. This corresponds to the idea that all selective processes operate not on an individual per se but on the individual's interaction with its environment (no individual is an island—and no fitness effect is visible unless you pair it with food, water, shelter, predators, members of the opposite sex, etc.). The effects of a single environmental variable (physical or biological) on a single individual have proven very insightful in dissecting mechanisms that underlie a specific behavior. Analogous to the current "systems approach" to cell and molecular processes, we have begun to study the biology of the fly at a group level using network theory. Network analysis may be used to study individuals and their interaction patterns within a group. This analysis assumes that the probability of one individual interacting with another, and how they interact, is influenced by prior interaction partners, prior interaction experiences, and the current social context—conditions met by _Drosophila_.

Drosophila has several well-studied, well-defined social behaviors that could be studied as network interactions. These social interactions include copulation, courtship, and aggressive bouts. Within a group, most interactions between flies share three fundamental characteristics: First, the _orientation_ between individuals is meaningful to the type of interactions; we quantify courtship duration and aggressive outcomes by scoring orientation. Second, the quality of these social interactions often depends on _proximity_; lunging, licking, tapping, boxing all require a minimum distance between flies. Finally, these behaviors operate for an extended _duration_; they are not simply "encounters" between passing flies.

To build a "foundation" of organization that includes any and all of these well-defined social behaviors, their spatial and temporal properties would have to be generalized into simple rules. Our recent study (Schneider et al., 2012) applied this method to determine the quality of interaction between flies that satisfy three criteria: (i) a criterion of orientation between a fly's heading and his interacting partner, (ii) a criterion such that they must be at a minimum distance, and (iii) these two criteria occur for a specified duration. Now this is not to discount potentially important interactions that occur for shorter durations (such as a lunge), or across longer distances (pheromonal communication), nor does it place special significance on the orientation—a female orientating away from a male is arguably as much an interaction as her orienting toward him. However, the quality of these many types of interactions is different. It is only a defined subset of these that is captured by these interactive criteria, and it is this limited subset that give rise to our Social Interaction Networks (SINs).

We were able to demonstrate that wild-type strains of _Drosophila_ form nonrandom SINs, and that these appear to have specific structural properties. We found that betweenness centrality (Newman, 2010) is strain dependent. Betweenness centrality is an index of network cohesion and information relay (for details, see Schneider et al., 2012). These observations indicate that an individual's importance in maintaining SIN connectivity varies with strain. We note that this structural property is one that has been found to be heritable in other species who exhibit social networks (Fowler et al., 2009). This nonrandom, strain-specific SIN organization among flies in a group belies the notion that _Drosophila_ mindlessly aggregate together and regulate their choice of interacting partner on an individual–individual basis based on chance and proximity. Rather, this choice of with whom to interact seems to be at least partially dependent on an individual's position in the SIN—a highly connected fly has a proportionally higher chance of receiving an interaction (also known as "preferential attachment"—and also a property of many social networks

(Newman, 2001)). This preferential attachment appears to be a dynamic process; highly interactive flies do not stay highly interactive for extended periods of time. Nevertheless, for short periods, flies seem to be able to indicate their degree of connectedness within a SIN (through behavior and/or chemical communication), and this may play a critical part in network organization.

To investigate channels of communication that mediate interactions in the network (both general communication as well as preferential attachment), we analyzed networks formed by fly strains harboring genetic mutations that produce severe disruptions of the sensory modalities (Schneider et al., 2012). Visual or auditory deficiencies did not impact networks significantly, while impairment of the gustatory receptors created networks that were not significantly different from random, and impairment of olfactory receptors severely altered network structure. We also observed that despite a lack of aggression or courtship behavior, flies frequently engaged in touching one another using both forelegs and middle legs. Combined with the aforementioned effect of gustation on network structure, the frequency and spatial patterning of touch suggest an underappreciated role for somatosensory communication.

One remarkable aspect of SINs in wild-type and mutant strains is that they exhibit consistent structural properties over time (Schneider et al., 2012). Considering that any one individual does not maintain a consistent pattern of connectivity, this SIN stability is somewhat paradoxical at the individual level. The rules underlying the interactions may therefore be dynamic properties of the group; each individual's probability to initiate or receive an interaction with another fly may follow from a complex evaluation of recent interactions and the social context of all other individuals. These rules, operating with gustatory and olfactory inputs and with strain-specific effects, are tantalizing in their suggestion of genetic regulation and coding for the expression of group patterns of connectivity.

Looking at patterns of conspecific social interaction, we can see that they follow certain rules, and the underpinnings of the rules are somehow regulated by genetic variation (Schneider et al., 2012). Yet, this is not a quantification of the social organization; this is the start of a foundation, a proof of strain-specific regulation in a homogeneous environment quantifying homogeneous interactions. One could readily expand this assay to a heterogeneous social environment (either via groups of mixed strains or species or via mixed sexes) or a heterogeneous physical environment and then create several networks layered on top of each other. Basic interaction patterns such as courtship, aggression, copulation, territoriality, and oviposition are not isolated, and they are not insulated from each other. We speculate that these individual interactions at the dyadic level lead to the group level where they are linked. How would these different networks correlate? What are their shared structures? Do they rely on common sensory modalities identified at the dyadic level? Do they rely on common integrative methods? Are they somewhat independent, where one pattern can exist without the other? These questions are becoming tractable and experiments will start to answer whether networks are simply an expression of some primitive organizational instinct or whether individual networks are competitively jostling for expression based on the relative fitness value of the behavioral patterns in question.

This SIN paradigm for studying social behaviors at the group level shows promise, not only for its ability to describe and quantify "foundational" networks but also for its applicability in studying more operational networks such as those formed by aggressive interactions or copulation. We also speculate that understanding rules that produce the patterns and the patterns behind group structure will ultimately allow us to understand how emergent phenotypes can arise from groups of *Drosophila*.

Network analysis also offers a nuanced approach to the study of genetic, cellular, and molecular mechanisms. It has been used for protein–protein interaction networks, neuronal networks, metabolic networks, behavioral patterns of interaction; at each stage, we see a dynamic process shaped by the interactions within. This ongoing formation implies that a "static" network would be the exception, and any phenomena with responsiveness finds itself embedded in a dynamic network created via the interactions between it and its neighbors. This paradigm of constant modulation via interactions is increasingly encountered at all levels of biology; even chromosomal regions undergo movement, joining and separating various regions, creating interactions (e.g., transcription factories (Osborne *et al.*, 2004)). With each level, we observe that networks are never constant in their interactions—transcription factors are not always bound to DNA, proteins are not always active, individuals are not always interacting—they are separated by time and space. Yet, the properties of networks are often remarkably constant (both SINs and regulatory networks (Milo *et al.*, 2004; Schneider *et al.*, 2012)), and at all levels—from gene to organism—emergent properties, patterns of association, and communication of information may be seen.

Although there is no current method of mapping dynamic relationships across network "levels," the use of networks to evaluate the pattern of interactions over a period of time, combined with the ability to monitor the evolution (or maintenance) of network properties, offers a powerful tool in the analysis of relationships among interactors. In this way, we can begin to decipher the complex systems at any level. Eventually, we will be able to tie networks at the organism level with ever more precise networks at the cell, gene, and protein level. And while this approach may initially draw upon the wealth of detail at the proteonomic level, it also offers promise at identifying and investigating the nonlinear pathways of genes whose expression seems to be influenced by social interactions (e.g., *aru* (Eddison *et al.*, 2011)). Looking at the structure, sequence, SNPs, promoters, and enhancers for a gene now goes hand in hand with expression patterns, domains, interactions, and regulatory functions for a protein, examining the gene in the context of the system. Hopefully, we can begin to integrate the social aspect, the "group"-level interaction into the system-level understanding of a gene, and in this way truly begin the genetic analysis of rules and patterns of social behavior, and the group phenotypes that they create.

IV. CONCLUSION

The complexity and challenge of evaluating group-influenced behavior arises because observing a specific behavior necessarily involves consideration of other behaviors—how does locomotion and aggression influence sleep, for example? Within a group, behavior of an individual influenced by his social environment feeds back onto the group, which in turn might readjust its influence. Once this concept is applied to the variety of behaviors that are known to be influenced by the social context, many of which occur simultaneously, we begin to see that groups of *Drosophila* can be considered complex systems with many interactions and feedback loops between individuals. The group is a complex system that may give rise to the emergence of group-level behavioral patterns.

The discovery of emergent group-level phenomena also shows that a group is a complex system with phenotypes that could not have been accurately predicted by simply "scaling-up" dyadic interactions. Rather, robust emergent properties are seen. This is intuitive as these behaviors did not evolve in isolation; they occur and were selected within, and interact with, the social environment. In other words, we will not be able to really understand the neurobiology of behaviors such as courtship, mating, and aggression until we can account for the role of the group.

When a phenotype (including a group-level one) is robust and depends on a genetic component, the question becomes "how?" What is happening in the group that facilitates ARM retrieval? How does the "group" explain as much variation in behavior as the genetic component? How does the mixing of two strains of males produce a remarkable fixed split in the paternity of the offspring independent of the ratio of the males? How does this pattern of offspring diversity arise even when any individual female does not produce this paternity ratio? We are forced to conclude that certain properties of the group are not simply properties of individuals or dyads "scaled-up" in number.

How does an individual's genotype contribute to emergent phenomena at the group level? How does the group interact with an individual's genetic composition? We are reminded of the similar paradox discussed by Bate; whereas he discussed the genetic contribution to neurons that interacted and associated to create emergent, organism-level behavior (Bate, 1998), we consider the genetic contribution to individuals, associating and interacting in a social space to create group-level phenomena. In exploring this paradigm, we come to the question, as did Bate, of how is it possible that the complexity of the group can be coded in the genetic composition of the individual? Is the complexity a "global dynamical system with many interactions" or rather "defined subprograms that [organisms] can get hold of and execute for themselves" (Brenner 1974)?

Given the potential complexity of the social environment of *Drosophila* in the wild (variations in sex ratios, number of conspecifics, amount, and number of various hetero-specifics (Shorrocks, 1972)), it becomes improbable that flies have their behavioral outputs (social "roles" or responses) hard-coded to respond to their social context. This suggests that the interactions throughout the group regulate individual behavior in a dynamic fashion to form a cohesive group effect.

These "interactions" in theory are any and all interactions between one or more flies. However, one could hypothesize that interactions which feedback to regulate social behavior would themselves be affected by social context, allowing a much more dynamic process of group interaction. This modulation of robust, innate behaviors could potentially explain the propagation of effects and stabilization of a group: each fly's social behavior has been modulated by its own history and is being modulated by its surroundings. In a dynamic system, exploring a single propagation of effect among many seems daunting. In theory, one should be able to attempt to discern the patterns that result from the behavioral interactions, and how these are organized to establish group cohesion (cohesion in the sense required for a robust group-level effect).

The ability to understand the group, therefore, appears to be linked with the ability not only to understand how each behavior is modulated by social context (of which much has been shown) but also to understand how these behaviors are patterned throughout the group, how these complex patterns arise from individuals, and eventually how this information is contained in the genome. One approach is to study the social group on a systems-level basis via network analysis. This is not a novel approach to group dynamics and has a long history of successful insights into the social organization of humans, monkeys, dolphins, and ants among others (Croft *et al.*, 2008; Formica *et al.* 2011).

Hopefully, the genetic dissection of patterns that may underlie emergent group phenotypes will help us understand the complexity of a group. However, this network investigation cannot exist without continued experimentation of social behaviors in other contexts. Dyadic interactions currently allow greater detail and throughput, and offer promise to continue to shed light on the implications of social interactions for the individual, yet they are simply one part of the continuum of social and groups. We are not simply asking about one versus two versus many, we are asking about one *among* many and many *among* more. Using a multilevel approach across disciplines, we may be able to understand the social life in more-than-minimal groups of *D. melanogaster*, the insect formerly known as "solitary."

References

Allada, R., and Chung, B. Y. (2011). Circadian organization of behavior and physiology in Drosophila. *Annu. Rev. Physiol.* **72,** 605–624.

Bate, M. (1998). Making sense of behavior. *Int. J. Dev. Biol.* **42**(3), 507–509.

Battesti, M., Moreno, C., Joly, D., and Mery, F. (2012). Spread of social information and dynamics of social transmission within Drosophila groups. *Curr. Biol.* **22**(4), 309–313.

Benzer, S. (1971). From the gene to behavior. *JAMA* **218**(7), 1015–1022.

Billeter, J. C., Atallah, J., Krupp, J. J., Millar, J. G., and Levine, J. D. (2009). Specialized cells tag sexual and species identity in Drosophila melanogaster. *Nature* **461**(7266), 987–991.

Billeter, J. C., Jagadeesh, S., Stepek, N., Azanchi, R., and Levine, J. D. (2012). Drosophila melanogaster females change mating behaviour and offspring production based on social context. *Proc. Biol. Sci.10.1098/rspb.2011.2676.*

Brenner, S. (1974). The genetics of Caenorhabditis elegans. *Genetics* **77**(1), 71–94.

Bretman, A., Fricke, C., and Chapman, T. (2009). Plastic responses of male Drosophila melanogaster to the level of sperm competition increase male reproductive fitness. *Proc. Biol. Sci.* **276**(1662), 1705–1711.

Chabaud, M. A., Isabel, G., Kaiser, L., and Preat, T. (2009). Social facilitation of long-lasting memory retrieval in Drosophila. *Curr. Biol.* **19**(19), 1654–1659.

Chan, Y. B., and Kravitz, E. A. (2007). Specific subgroups of FruM neurons control sexually dimorphic patterns of aggression in Drosophila melanogaster. *Proc. Natl. Acad. Sci. U.S.A.* **104** (49), 19577–19582.

Chen, S., Lee, A. Y., Bowens, N. M., Huber, R., and Kravitz, E. A. (2002). Fighting fruit flies: A model system for the study of aggression. *Proc. Natl. Acad. Sci. U.S.A.* **99**(8), 5664–5668.

Croft, D. P., James, R., and Krause, J. (2008). Exploring Animal Social Networks. Princeton University Press, Princeton, Oxford.

Davis, R. L. (2004). Olfactory learning. *Neuron* **44**(1), 31–48.

Dierick, H. A. (2008). Fly fighting: Octopamine modulates aggression. *Curr. Biol.* **18**(4), R161–R163.

Dierick, H. A., and Greenspan, R. J. (2006). Molecular analysis of flies selected for aggressive behavior. *Nat. Genet.* **38**(9), 1023–1031.

Dierick, H. A., and Greenspan, R. J. (2007). Serotonin and neuropeptide F have opposite modulatory effects on fly aggression. *Nat. Genet.* **39**(5), 678–682.

Donlea, J. M., Ramanan, N., and Shaw, P. J. (2009). Use-dependent plasticity in clock neurons regulates sleep need in Drosophila. *Science* **324**(5923), 105–108.

Dow, M. A., and von Schilcher, F. (1975). Aggression and mating success in Drosophila melanogaster. *Nature* **254**(5500), 511–512.

Eddison, M., Guarnieri, D. J., Cheng, L., Liu, C. H., Moffat, K. G., Davis, G., and Heberlein, U. (2011). Arouser reveals a role for synapse number in the regulation of ethanol sensitivity. *Neuron* **70**(5), 979–990.

Ehrman, L. (1970). The mating advantage of rare males in Drosophila. *Proc. Natl. Acad. Sci. U.S.A.* **65**(2), 345–348.

Ehrman, L. (1972). A factor influencing the rare male mating advantage in Drosophila. *Behav. Genet.* **2**(1), 69–78.

Formica, V. A., Wood, C. W., Larsen, R. E., Butterfield, M. E., Augat, M. E., Hougen, H. Y., and Brodie, E. D., III (2011). Fitness consequences of social network position in a wild population of forked fungus beetles (Bolitotherus cornutus). *J. Evol. Biol.* **25**, 130–137.

Fowler, J. H., Dawes, C. T., and Christakis, N. A. (2009). Model of genetic variation in human social networks. *Proc. Natl. Acad. Sci. U.S.A.* **106**(6), 1720–1724.

Fujii, S., Krishnan, P., Hardin, P., and Amrein, H. (2007). Nocturnal male sex drive in Drosophila. *Curr. Biol.* **17**(3), 244–251.

Ganguly-Fitzgerald, I., Donlea, J., and Shaw, P. J. (2006). Waking experience affects sleep need in Drosophila. *Science* **313**(5794), 1775–1781.

Hall, J. C. (2002). Courtship lite: A personal history of reproductive behavioral neurogenetics in Drosophila. *J. Neurogenet.* **16**(3), 135–163.

Higgins, L. A., Jones, K. M., and Wayne, M. L. (2005). Quantitative genetics of natural variation of behavior in Drosophila melanogaster: The possible role of the social environment on creating persistent patterns of group activity. *Evolution* **59**(7), 1529–1539.

Hoffmann, A. (1987a). A laboratory study of male territoriality in the sibling species Drosophila melanogaster and D. simulans. *Anim. Behav.* **35**(3), 807–818.

Hoffmann, A. (1987b). Territorial encounters between Drosophila males of different sizes. *Anim. Behav.* **35**(6), 1899–1901.

Hoffmann, A. (1988). Heritable variation for territorial success in two Drosophila melanogaster populations. *Anim. Behav.* **36**(4), 1180–1189.

Hoffmann, A. A. (1990). The influence of age and experience with conspecifics on territorial behavior in *Drosophila melanogaster*. *J. Insect Behav.* **3**(1), 1–12.

Kahsai, L., and Zars, T. (2011). Learning and memory in Drosophila: Behavior, genetics, and neural systems. *Int. Rev. Neurobiol.* **99**, 139–167.

Keene, A. C., and Waddell, S. (2007). Drosophila olfactory memory: Single genes to complex neural circuits. *Nat. Rev. Neurosci.* **8**(5), 341–354.

Kent, C., Azanchi, R., Smith, B., Formosa, A., and Levine, J. D. (2008). Social context influences chemical communication in D. melanogaster males. *Curr. Biol.* **18**(18), 1384–1389.

Kravitz, E. A., and Huber, R. (2003). Aggression in invertebrates. *Curr. Opin. Neurobiol.* **13**(6), 736–743.

Krupp, J. J., Kent, C., Billeter, J. C., Azanchi, R., So, A. K., Schonfeld, J. A., Smith, B. P., Lucas, C., and Levine, J. D. (2008). Social experience modifies pheromone expression and mating behavior in male Drosophila melanogaster. *Curr. Biol.* **18**(18), 1373–1383.

Lefranc, A., Jeune, B., Thomas-Orillard, M., and Danchin, E. (2001). Non-independence of individuals in a population of Drosophila melanogaster: Effects on spatial distribution and dispersal. *C. R. Acad. Sci. III* **324**(3), 219–227.

Levine, J. D., Funes, P., Dowse, H. B., and Hall, J. C. (2002). Resetting the circadian clock by social experience in Drosophila melanogaster. *Science* **298**(5600), 2010–2012.

Lone, S. R., and Sharma, V. K. (2011). Social synchronization of circadian locomotor activity rhythm in the fruit fly Drosophila melanogaster. *J. Exp. Biol.* **214**(22), 3742–3750.

Lupold, S., Manier, M. K., Ala-Honkola, O., Belote, J. M., and Pitnick, S. (2011). Male Drosophila melanogaster adjust ejaculate size based on female mating status, fecundity, and age. *Behav. Ecol.* **22**(1), 184–191.

McGuire, S. E., Deshazer, M., and Davis, R. L. (2005). Thirty years of olfactory learning and memory research in Drosophila melanogaster. *Prog. Neurobiol.* **76**(5), 328–347.

Mery, F., Varela, S. A., Danchin, E., Blanchet, S., Parejo, D., Coolen, I., and Wagner, R. H. (2009). Public versus personal information for mate copying in an invertebrate. *Curr. Biol.* **19**(9), 730–734.

Miczek, K. A., de Almeida, R. M. M., Kravitz, E. A., Rissman, E. F., de Boer, S. F., and Raine, A. (2007). Neurobiology of escalated aggression and violence. *J. Neurosci.* **27**(44), 11803–11806.

Milo, R., Itzkovitz, S., Kashtan, N., Levitt, R., Shen-Orr, S., Ayzenshtat, I., Sheffer, M., and Alon, U. (2004). Superfamilies of evolved and designed networks. *Science* **303**(5663), 1538–1542.

Moore, A. J., Brodie, E. D., III, and Wolf, J. B. (1997). Interacting phenotypes and the evolutionary process: I direct and indirect genetic effects of social interactions. *Evolution.* **51**(5), 1352–1362.

Newman, M. E. (2001). Clustering and preferential attachment in growing networks. *Phys. Rev. E Stat. Nonlin. Soft Matter Phys.* **64**(2 Pt 2), 025102.

Newman, M. (2010). Networks: An Introduction. Oxford University Press, Oxford.

Osborne, C. S., Chakalova, L., Brown, K. E., Carter, D., Horton, A., Debrand, E., Goyenechea, B., Mitchell, J. A., Lopes, S., Reik, W., *et al.* (2004). Active genes dynamically colocalize to shared sites of ongoing transcription. *Nat. Genet.* **36**(10), 1065–1071.

Robin, C., Daborn, P. J., and Hoffmann, A. A. (2007). Fighting fly genes. *Trends Genet.* **23**(2), 51–54.

Sakai, T., Isono, K., Tomaru, M., Fukatami, A., and Oguma, Y. (2002). Light wavelength dependency of mating activity in the Drosophila melanogaster species subgroup. *Genes Genet. Syst.* **77**(3), 187–195.

Saltz, J. B., and Foley, B. R. (2011). Natural genetic variation in social niche construction: Social effects of aggression drive disruptive sexual selection in Drosophila melanogaster. *Am. Nat.* **177** (5), 645–654.

Sarin, S., and Dukas, R. (2009). Social learning about egg-laying substrates in fruitflies. *Proc. Biol. Sci.* **276**(1677), 4323–4328.

Schneider, J., Dickinson, M. D., and Levine, J. D. (2012). Biological embedding of early social adversity: From fruit flies to kindergartners sackler colloquium: Social structures depend on innate determinants and chemosensory processing in *Drosophila*. PNAS; published ahead of print July 16, 2012, http://dx.doi.org/doi:10.1073/pnas.1121252109.

Shorrocks, B. (1972). Drosophila. Ginn, London.

Singh, S. R., Singh, B. N., and Hoenigsberg, H. F. (2002). Female remating, sperm competition and sexual selection in Drosophila. *Genet. Mol. Res.* **1**(3), 178–215.

Suh, G. S., Wong, A. M., Hergarden, A. C., Wang, J. W., Simon, A. F., Benzer, S., Axel, R., and Anderson, D. J. (2004). A single population of olfactory sensory neurons mediates an innate avoidance behaviour in Drosophila. *Nature* **431**(7010), 854–859.

Tauber, E., Roe, H., Costa, R., Hennessy, J. M., and Kyriacou, C. P. (2003). Temporal mating isolation driven by a behavioral gene in Drosophila. *Curr. Biol.* **13**(2), 140–145.

Wang, L., and Anderson, D. J. (2010). Identification of an aggression-promoting pheromone and its receptor neurons in Drosophila. *Nature* **463**(7278), 227–231.

Wertheim, B., van Baalen, E. J., Dicke, M., and Vet, L. E. (2005). Pheromone-mediated aggregation in nonsocial arthropods: An evolutionary ecological perspective. *Annu. Rev. Entomol.* **50**, 321–346.

Yurkovic, A., Wang, O., Basu, A. C., and Kravitz, E. A. (2006). Learning and memory associated with aggression in Drosophila melanogaster. *Proc. Natl. Acad. Sci. U.S.A.* **103**(46), 17519–17524.

4

The Circadian Clock of the Fly: A Neurogenetics Journey Through Time

Özge Özkaya and Ezio Rosato

Department of Genetics, University of Leicester, Leicester, United Kingdom

Advances in Genetics, Vol. 77   0065-2660/12 $35.00
Copyright 2012, Elsevier Inc. All rights reserved.   http://dx.doi.org/10.1016/B978-0-12-387687-4.00004-0

ABSTRACT

Forty years ago, a mutagenesis screening in the fruit fly, *Drosophila melanogaster*, led to the discovery of *period*, the first gene to be involved in the endogenous 24-h rhythmicity of an organism. Since then circadian clocks have been identified in fungi, cyanobacteria, plants, and other animals. Although the molecular components are not conserved across the main divisions of life, it appears that in every organism, a common design, based upon a transcription–translation feedback loop (TTL), is in place to regulate endogenous 24 h cycles. The TTL model has informed chronobiology research for the majority of the past 30 years with spectacular results. However, new evidence and the rediscovery of old observations suggest that this model is coming to age. Here, we provide a comprehensive review of the current TTL model in *Drosophila* highlighting its accomplishments and its limitations. We conclude by offering our personal view on the organization and the evolution of circadian clocks.   © 2012, Elsevier Inc.

I. INTRODUCTION

Life is constantly under the scrutiny of natural selection. Individuals are confronted with the need of finding food, shelter, and mates, and competition is high. The 24 h changes in abiotic (light and temperature) and biotic (social interactions) factors are main determinants of survival and reproduction. Therefore, it is not surprising that an endogenous, ~24-h, (circadian) clock has evolved to coordinate the activities of an organism with the 24 h cycles of its environment, resulting in a measurable increment in fitness (Ouyang *et al.*, 1998). About 40 years ago, a screen to identify genes involved in the circadian regulation of the rest-activity cycle of the fruit fly led, against all odds, to the discovery of the first circadian (and behavioral) gene, *period* (*per*) in *Drosophila melanogaster* (Konopka and Benzer, 1971). Incidentally, this event marked the beginning of modern neurogenetics, the study of how genes organize the structure and dictate the function of the nervous system. *Drosophila* is the perfect model system to study neurogenetics, as the complexity of its behavioral repertoire contrasts with the simplicity of its genetic manipulation. Four decades of genetics and molecular biology applied to the dissection of the circadian clock have generated a good (albeit incomplete) understanding of the functioning of several interdependent clock genes that are a fundamental component of the time-keeping mechanism. The discovery and description of these genes, whose protein products beautifully interact one with another in a 24-h coordinated fashion, have led to the idea that circadian timing relies on cell-autonomous regulation of rhythmic gene expression. In other words, it is hypothesized that all circadian phenomena, such as cycles of biochemical reactions, changes in physiology, or rhythmic behavior, depend upon clock gene expressing cells, and that clock cells are individually able to sustain their own rhythmicity. This paradigm of circadian function has become known as the negative transcription/translation feedback loop (TTL) model, where key high-order regulatory genes achieve rhythmic transcription by controlling their own expression and then transmit oscillatory information to a second tier of genes, called clock controlled genes (ccgs) (Dunlap, 1999). These are not part of the core of the clock *per se*; instead, they set into motion cell/tissue specific rhythmic outputs. An essential feature of the model is that several clock proteins act as inhibitors of transcription. They are ineffective at first but then become competent negative regulators after progressive posttranslational modifications that also mark the end of their cycle, thus explaining the necessary temporal delay for oscillatory changes to occur. Interestingly, genomic analyses suggest that the clock takes overall control of gene expression, as microarray studies have shown a substantial part of the transcriptome cycles with a 24-h rhythm in complex tissues (Ceriani *et al.*, 2002; Claridge-Chang *et al.*, 2001; McDonald and Rosbash, 2001; Ueda *et al.*, 2002) and in small, more homogenous groups of neurons (Kula-Eversole *et al.*, 2010; Nagoshi *et al.*, 2010).

     Not all clock cells are born equal. The traditional view is that there are "main" circadian or "central" clock cells that are seemingly able to endure prolonged periods of constant conditions and "peripheral" clock cells that apparently have a less robust clock. It is now clear that the negative feedback loop does not come in just one flavor, and that the distinction between central and peripheral clock cells is very different from what previously thought (see section V). Molecular components that experimental evidence portrays as essential for overt rhythmicity or robustness are not absolutely ubiquitous in every clock cell (Gummadova *et al.*, 2009; Lamaze *et al.*, 2011; Lim *et al.*, 2011; Yu *et al.*, 2011), suggesting that a degree of divergence might be the norm to fine tune the feedback loop to the needs of different clock neurons. This view is in line and complements more recent observations that several types of neurons are necessary to interpret and react to different rhythmic environmental conditions in an appropriate fashion (for instance, long or short photoperiod, lower or higher temperature, etc.,), even in a highly simplified, laboratory-based version of the world (Busza *et al.*, 2007; Grima *et al.*, 2004; Stoleru *et al.*, 2004, 2005, 2007). Moreover, seminal experiments have shown that membrane activity (Nitabach *et al.*, 2002, 2006; Sheeba *et al.*, 2008a,b) and intercellular communication (Grima *et al.*, 2004; Stoleru *et al.*, 2004, 2005, 2007) are essential for robust cycling behavior. Taken to the limit, these observations suggest that circadian rhythmicity is not the additive product of individual neurons but the emerging property of a network (Harmar *et al.*, 2002; Lin *et al.*, 2004; Maywood *et al.*, 2006; Nitabach *et al.*, 2006; Peng *et al.*, 2003; Yamaguchi *et al.*, 2003), an hypothesis that contrasts with aspects of the cell-autonomous concept of the clock and requires an update of the negative feedback loop model.

Another exciting but disconcerting idea is the spreading notion that TTLs might not represent the essence of the clock, but mechanisms to sustain and enhance an "enzymatic" clock able to impart rhythmicity to cell physiology in the absence of transcription and translation. This view, based on new findings (Nakajima *et al.*, 2005; O'Neill and Reddy, 2011; O'Neill *et al.*, 2011; Tomita *et al.*, 2005) and on the rediscovery of old results (Lakin-Thomas, 2006; Sweeney and Haxo, 1961), is challenging our elegant but simplistic understanding of the clock. The advance of the field will require a new vision able to incorporate our current genetic knowledge with the new physiological results. It will entail the discovery of the motors of the enzymatic clock, the description of how the latter connects to the components of the TTL and the understanding of the logic behind a multicomponent and multicellular clock that is more powerful than the sum of its parts. We will explore further these concepts at the end of the chapter, but we shall first present an overview of the *Drosophila* clock as seen by our current, orthodox prospective. Although this review deals with the *Drosophila* clock, there will be some reference to the clock of other organisms to point to important questions on clock mechanisms that transcend the system under study.

II. THE MOLECULAR COMPONENTS OF THE CLOCK

At the molecular level, our current understanding of the clock is based upon two interlocked negative TTLs. The basic helix–loop–helix (bHLH) and PAS domain transcriptional activators CLOCK (CLK) and CYCLE (CYC) (Allada *et al.*, 1998; Rutila *et al.*, 1998), sit at the core of the two loops. They hetero-dimerize via their HLH and PAS domains and bind to the circadian E-box sequence CACGTG, initiating transcription. E-boxes are found on the promoters of the so-called evening transcripts genes, such as *period* (*per*), *timeless* (*tim*), *vrille* (*vri*), and *Par domain protein 1ε* (*Pdp1ε*) (Allada *et al.*, 1998; Cyran *et al.*, 2003; Glossop *et al.*, 1999, 2003; Rutila *et al.*, 1998) (Fig. 4.1).

A. The first loop (part I): PER and TIM accumulation results in the late night/early morning repression of the "evening transcripts"

The evening transcripts peak during the first few hours of darkness. At that time, CLK and CYC are stably anchored onto the E-boxes, CLK is hypophosphory-lated, and PER and TIM are virtually absent. This is because the increase in PER and TIM amounts is delayed by about 6 h (Edery *et al.*, 1994; Hardin *et al.*, 1992; Myers *et al.*, 1995) due to destabilizing posttranslational modifications, phosphorylation in particular (see below). In the middle of the night, accumulating PER–TIM dimers enter the nucleus (Saez and Young, 1996; Vosshall *et al.*, 1994; Zerr *et al.*, 1990) and start repressing the CLK–CYC complex (Lee *et al.*, 1999).

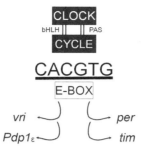

Figure 4.1. CLOCK (CLK) and CYCLE (CYC) drive transcription of evening genes. CLK and CYC are bHLH–PAS transcription factors. They dimerize through their HLH and PAS domains and through the basic portion of the bHLH region bind to the sequence CACGTG, called an E-box. E-boxes are present in the promoter region of "evening transcript" genes such as *period (per)*, *timeless (tim)*, *vrille (vri)*, and *Par domain protein 1ε (Pdp1ε)*. (See Color Insert.)

From that point onwards, hyperphosphorylated forms of CLK become increasingly prevalent, ultimately leading to a halt in the transcription of the evening genes (Lee *et al.*, 1998) (Fig. 4.2).

B. The second loop (part I): VRI and PDP1ε time the expression of the "morning transcripts"

The functioning of the second loop is much less clear. Experimental data have shown that transcripts of *Clk* and *cryptochrome* (*cry*, another circadian gene involved in light entrainment in *Drosophila*—see below) peak in the late night/early morning (Allada *et al.*, 1998; Emery *et al.*, 1998), at times when repression of CLK–CYC is at its maximum. This means that additional positive and negative transcription regulators must be implicated, operating in antiphase with CLK–CYC and PER–TIM, respectively. In wild-type flies, the peak of VRI expression, a basic leucine zipper (bZIP) protein, quickly follows that of its RNA, coinciding with times when *Clk* transcripts are at their lowest. Moreover, overexpression of VRI stops the clock, resulting in low levels of *Clk* mRNA and protein (Blau and Young, 1999). This and other evidence suggest that VRI, after forming homodimers that bind to the TTATGTAA sequence (called a V/P box) found in the promoter of morning genes, is able to block transcription (Cyran *et al.*, 2003; Glossop *et al.*, 2003). If VRI is the repressor, the activator must then be another bZIP protein able to compete with VRI. The activator should bind to V/P boxes at a later phase of the cycle, resulting in the expression of the morning transcripts. In mammals, the V/P box is recognized by rhythmic

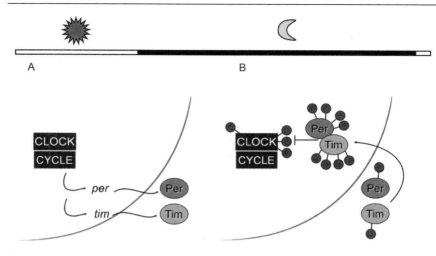

Figure 4.2. PER and TIM feedback on CLK and CYC to inhibit the transcription of evening genes. (A) Transcription of *per* and *tim* starts at the end of the day reaching a peak at the beginning of the night. PER and TIM proteins accumulate with a delay of about 6 h, compared to their transcripts, due to progressive phosphorylation events that regulate their stability. (B) By the end of the night, the PER–TIM dimer, extensively phosphorylated, has accumulated into the nucleus where it interacts with the CLK–CYC complex. This interaction causes the phosphorylation of CLK, which results in the repression of transcription. The sun and moon symbols and the white–black–white bars represent the day–night cycle. (See Color Insert.)

bZIP PAR domain proteins (Lopez-Molina *et al.*, 1997; Mitsui *et al.*, 2001), suggesting that an analogous protein might be the activator in flies. Indeed, PDP1ε is a nuclear protein that can form homodimers able to bind to the V/P boxes (Cyran *et al.*, 2003), and the isoform ε (the only one out of six) of the *Pdp 1* gene is the only PAR domain protein encoding mRNA to cycle in flies heads (McDonald and Rosbash, 2001; Ueda *et al.*, 2002). Both *Pdp1ε* and *vri* mRNAs accumulate in unison, but their protein products do not, as PDP1ε lags VRI accumulation by about 3–6 h (Cyran *et al.*, 2003). This observation suggests a model by which both genes are transcribed by CLK–CYC in the late day–early night. At the same time, VRI, which accumulates quickly, binds to V/P boxes in the promoter region of morning genes such as *Clk* and *cry*, inhibiting transcription. After 3–6 h, the accumulation of PDP1ε releases the inhibition allowing transcription to occur in the late night–early day (Cyran *et al.*, 2003) (Fig. 4.3). Since CLK drives the expression of *vri* and *Pdp1ε*, the rhythm of *Clk* expression constitute a second negative feedback loop that is interconnected to the first (the *per/tim* loop), by virtue of sharing common elements (Fig. 4.4).

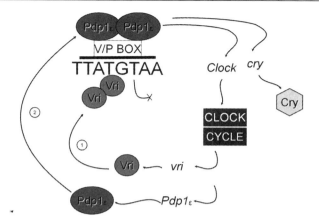

Figure 4.3. Regulation of *Clk* transcription. The positive factors initiating the transcription of *Clk* are currently not identified. However, the protein products of two evening transcripts, VRILLE (VRI) and PAR DOMAIN PROTEIN 1ε (PDP1ε), regulate CLK transcription. VRI accumulates in the early night ahead of PDP1ε, binds to the TTATGTAA sequence (V/P box), and inhibits *Clk* transcription. About 3–6 h later, in the late night, PDP1ε outcompetes VRI from the V/P binding sites allowing transcription of *Clk* to resume. (See Color Insert.)

C. The second loop (part II): an imperfect picture

The relevance of PDP1ε in the second loop has been challenged. Benito *et al.* (2007) overexpressed PDP1ε or downregulated (by RNAi) the level of all PDP1 isoforms specifically in a group of clock neurons, the PIGMENT DISPERSING FACTOR (PDF)-expressing s-LNv$_s$ (see below). Both manipulations rendered the flies arrhythmic in constant darkness (DD) but did not affect the molecular cycling of the cellular clock. Lim *et al.* (2007b) also reached similar conclusions using the overexpression of a dominant-negative form of PDP1. Thus, it was suggested that PDP1ε might regulate behavioral output rather than being important for circadian rhythms *per se* (Benito *et al.*, 2007; Lim *et al.*, 2007b). More recently, the discovery of a mutation affecting specifically the 1ε variant of *Pdp* showed that PDP1ε is indeed involved in the transcriptional activation of *Clk* (CLK levels were very low in the mutant) and in general in the correct regulation of the clock (in the mutant PER levels were also low, possibly explaining the additional hypophosphorylation of CLK) (Zheng *et al.*, 2009). However, these molecular effects were much more pronounced in the s-LNvs than in peripheral clock cells such as the eyes and in DD rather than in light–dark (LD) conditions. It follows that several additional components are likely to participate in the regulation of the second feedback loop, and that *Clk* transcription can be activated by different factors in different cell types. Indeed, *Clk*

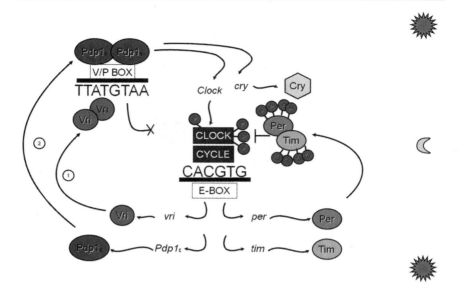

Figure 4.4. Two interlocked negative feedback loops. Rhythmic transcription of *per* and *tim* is determined by CLK–CYC, acting as positive elements and by PER–TIM that constitute the negative elements of the loop. *Clk* expression is regulated by VRI and PDP1ε, having a negative and positive effect on its transcription, respectively. As the two loops share common elements they are interlocked in their regulation. The sun–moon–sun sequence represents the day–night cycle. (See Color Insert.)

expression in different subgroups of central clock cells depends upon different regions on the *Clk* promoter, which does not necessarily contain V/P boxes (Gummadova *et al.*, 2009).

D. The first loop (part II): controversial regulators of evening genes

More recently, two additional transcription factors NEJIRE (NEJ) and CLOCK-WORK ORANGE (CWO) have been implicated in the expression of the evening genes. Both seem to modulate the function of the CLK–CYC complex, but their mode of action is not completely understood. NEJ is the *Drosophila* ortholog of the mammalian CREB BINDING PROTEIN (CBP). Since *cbp/nej* is an embryonic essential gene, loss of function studies could not be conducted. However, CBP/NEJ knockdown by RNAi in PDF-expressing cells (the LNvs) resulted in the lengthening of the period of locomotor activity in adult flies, with a parallel delay in the phase of rhythmic clock gene expression in LNvs and LNds (the latter not expressing the RNAi construct), as measured by immunofluorescence (Lim *et al.*, 2007a).

Conversely, CBP/NEJ overexpression in the LNvs had no apparent effect. Extending the overexpression to all the clock cells (*tim*-expressing cells) caused arrhythmic circadian behavior in DD and loss of the evening peak (see below) in LD conditions (Lim *et al.*, 2007a). Molecularly, CBP/NEJ overexpression resulted in lower PER levels. Interestingly, PER reduction in the PDF-positive LNvs was much more pronounced when CBP/NEJ overexpression was driven in all circadian cells, then when it was limited to the PDF-expressing neurons (Lim *et al.*, 2007a). This result correlates with the behavioral data, suggests that CBP/NEJ might have a different role in different clock cells and shows that cellular communication impacts on the expression of clock genes. A reduction in PER levels following CBP/NEJ overexpression might result from decreased transcription of the evening genes. To test this hypothesis, Lim *et al.* (2007a) overexpressed CBP/NEJ in S2 cells (an embryonic cell line of *Drosophila*) and detected a suppression in CLK–CYC-dependent transcription from reporter plasmids containing *per* and *tim* promoters (Lim *et al.*, 2007a). However, they also detected a moderate increase in the expression of a *per* reporter following cotransfection of *cbp/nej* and protein kinase A (*pka*). This can be explained considering that CBP/NEJ functions as a transcriptional coactivator of the CREB/PKA pathway (Kwok *et al.*, 1994), and that there are three putative CREB-binding sites between 4 and 1.2 kb upstream of the transcription start site for the *per* gene (Belvin *et al.*, 1999). Finally, overexpressed CBP/NEJ directly targeted the PAS A domain of CLK, inhibiting the formation of the CLK–CYC dimer in S2 cells (Lim *et al.*, 2007a) (Fig. 4.5).

In another study, Hung and colleagues also found that overexpression of CBP/NEJ resulted in behavioral arrhythmicity and decreased PER expression (Hung *et al.*, 2007). However, in cell culture, they obtained different results. Downregulating endogenous CBP/NEJ expression in S2 cells either by RNAi or by coexpression of the adenoviral E1A(12S) protein, a CBP inhibitor, decreased CLK–CYC-dependent expression from a four *per*-E-box containing reporter. This suggests that CBP/NEJ might actually be an activator of CLK–CYC transcription, in contrast to the results of Lim and coworkers. To prove this, Hung *et al.* (2007) assayed the level of the evening genes *tim* and *Pdp1ε* in wild-type and *cbp/nej* mutant flies under constant light (LL). Under such condition, the PER–TIM feedback loop, which compensates for changes in CLK–CYC activity, does not operate (because of degradation of TIM and PER, see below) allowing a more direct test of CLK–CYC functionality *in vivo*. Levels of both transcripts were reduced in a *cbp/nej* partial loss of function mutant compared to controls, as assayed by quantitative real-time reverse transcription PCR performed on head extracts (but note that the eyes constitute the most abundant component here). Conversely, levels of *tim* and *Pdp1ε* mRNA went up after heat shock in a heat-shock driven *cbp/nej* transgene but not in control flies. Those data indicate a coactivator function of CBP/NEJ for CLK–CYC-mediated transcription (Hung *et al.*, 2007). This is in contrast to the interpretation of Lim *et al.* (2007a) that

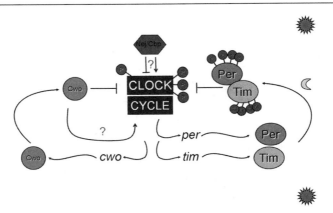

Figure 4.5. Controversial contribution of NEJ/CBP and CWO to the clock. NEJ/CBP interferes
with CLK–CYC-mediated transcription but it behaves as an activator or a repressor
according to the experimental protocol. CWO is a strong repressor of E-boxes containing promoters which it binds without dimerizing with either CLK or CYC. As *cwo*
transcription is activated by CLK–CYC and is repressed by its own protein product, *cwo*
regulation constitutes a third negative feedback loop interconnected with the others.
Experimental evidence suggests that early at night, when PER levels are low and
cytoplasmic, CWO might function as an activator of transcription, instead. Thus CWO
could have a dual clock function depending on the time of day. The sun–moon–sun
sequence represents the day–night cycle. (See Color Insert.)

CBP/NEJ might function as a negative regulator of the CLK–CYC heterodimer
(Fig. 4.5). Clearly, this issue needs revisiting, but considering the many differences between the two studies in terms of tissues examined, light conditions and
reagents utilized, it is possible that CBP/NEJ might exert both roles *in vivo*
according to cell type, environmental condition, and physiological state.

 A similar controversial conclusion has been reached by Richier *et al.*
(2008) regarding the second transcription factor CWO. CWO is a bHLH-
ORANGE protein that is rhythmically expressed under CLK–CYC control
(Kadener *et al.*, 2007; Lim *et al.*, 2007c; Matsumoto *et al.*, 2007; Richier
et al., 2008), enlisting it into the group of evening genes. Expression studies
using specific α-CWO antibodies or fly lines that trap *cwo* enhancers have
shown that *cwo* is expressed in all clock neurons (Kadener *et al.*, 2007; Lim
et al., 2007c; Matsumoto *et al.*, 2007; Richier *et al.*, 2008). Lack of CWO in
null mutants (Richier *et al.*, 2008), strong hypomorphs (Kadener *et al.*, 2007;
Lim *et al.*, 2007c), or via RNAi (Matsumoto *et al.*, 2007) resulted behaviorally
in a long period of locomotor activity and molecularly in altered expression of
the evening genes. In particular, a strong decrease in the peak and a modest
increase in the trough of evening genes mRNAs were generally observed,

coinciding with times when PER–TIM are at their lowest and highest level, respectively. Similar results were obtained both under LD and DD conditions. bHLH-ORANGE proteins are often DNA-binding transcriptional repressors, suggesting that CWO might repress CLK–CYC-driven transcription. Transcriptional assays were performed in S2 cells, showing that CWO is a strong repressor of E-boxes containing promoters (Kadener *et al.*, 2007; Lim *et al.*, 2007c; Matsumoto *et al.*, 2007), that binding to E-boxes does not require dimerization with either CLK (not expressed in S2) or CYC (*cyc* dsRNAs had not effect), and that PER cooperates with CWO to repress CLK–CYC-mediated transcription (Kadener *et al.*, 2007). Direct binding of CWO on circadian E-boxes was also confirmed in S2 cells, by chromatin immunoprecipitation followed by hybridization on a *Drosophila* genome tiling array (Chip-on-chip assay) (Matsumoto *et al.*, 2007). Taken together, these data have been interpreted as an indication that CWO functions as a dedicated repressor of CLK–CYC-driven transcripts and provides an additional mechanism to ensure the switching off of evening genes (Kadener *et al.*, 2007; Lim *et al.*, 2007c; Matsumoto *et al.*, 2007) (Fig. 4.5).

The reduction in the peak of mRNA generally seen for the evening genes has been interpreted as an indirect or "system" effect, since intermediate mRNA levels of CLK–CYC-dependent transcripts are seen in other repressor mutant strains, namely, per^{01} and tim^{01} (Kadener *et al.*, 2007). Furthermore, this explanation is in line with the observation that while the PER–TIM repressor is still functional in *cwo* loss of function mutants, the activator is impaired due to reduced levels and phosphorylation of CLK (Richier *et al.*, 2008). However, this simple explanation is at odds with the regulation of one of the evening genes, *cwo* itself. *cwo* mRNA levels were constitutively high in a *cwo*-null mutant, reaching about eight times the peak level in the wild type (Richier *et al.*, 2008). This suggests that CWO can be a strong repressor of transcription, and that it acts so on its own promoter. Conversely, CWO could have a dual function on the other E-boxes promoters. Considering the decrease in the peak and the increase in the trough of mRNA observed especially for *tim* (Richier *et al.*, 2008), and taking into account the cooperative effect of CWO and PER in achieving repression (Kadener *et al.*, 2007), Richier *et al.* (2008) suggest that CWO may function as a transcriptional activator in the evening when PER is low and cytoplasmic and as a repressor in the early morning when PER has accumulated in the cell nucleus (Fig. 4.5). To solve the controversy, further investigation is needed, but the emerging picture suggests that we are still quite far from a complete description of the negative feedback loops, that clock cells differ in the molecular details of their clock, and that cellular cross-talk (see above the description of CBP/NEJ) might extend the effect that a protein has on a few cells to the whole network.

E. Posttranscriptional regulation

In the previous paragraphs, we have discussed interactions among clock proteins resulting in rhythmic gene expression through regulation of transcription. However, as genetic information is first transcribed into mRNA and then translated into proteins, there are additional steps in the gene-to-protein cascade, which are amenable of rhythmic regulation. These are stability of transcripts, translational control, and posttranslational modifications. Indeed, circadian regulation of these mechanisms is necessary to generate significant fluctuations in the levels of clock proteins. Theoretically, for any mRNA produced by rhythmic transcription, it is expected a loss of amplitude and a delay in phase, compared to the amplitude and phase of transcription, which is more severe the longer the half-life of the RNA (Wuarin *et al.*, 1992). The same applies to proteins that are synthesized from cycling RNAs (Wood, 1995). To address this decrease in amplitude and to make sure that the protein products of rhythmically expressed genes reach high concentration just in anticipation of cellular needs, additional regulatory processes are needed.

1. mRNA stability

In the case of the *per* gene, a transcribed region carries a sequence that shortens *per* mRNA half-life. Stanewsky and coworkers compared the mRNA levels of two luciferase reporters, one called plo-luc, containing the promoter region of *per* but no coding sequence, the other, BG-luc, containing a genomic sequences encoding for the N-terminal two-thirds of PER in addition to the promoter region (Stanewsky *et al.*, 1997). The amplitude for plo-luc mRNA cycling was approximately three times lower than for BG-luc mRNA, the latter instead showed the same amplitude as for the endogenous *per* transcript. Moreover, in a *per*-null background (both reporters are unable to provide wild-type PER function), plo-luc mRNA was constitutively high whereas BG-luc mRNA, like endogenous *per*, reached intermediate levels. Both results are consistent with a longer half-life of plo-luc mRNA, suggesting the presence in the *per* transcript of a destabilizing element. A shorter half-life should result in an earlier phase of mRNA accumulation for BG-luc and *per* compared to plo-luc. Instead, the opposite was observed, with plo-luc mRNA accumulating about 2 h earlier than BG-luc and *per*. Therefore, this unidentified element must carry a dual regulatory function, as it delays *per* transcription in addition to destabilizing its own mRNA (Stanewsky *et al.*, 1997).

2. Control of translation

Factors controlling the translation of clock proteins have not been identified until very recently. Lim and coworkers discovered a protein, TWENTY-FOUR (TYF) that is required for robust PER translation in the LNvs (Lim *et al.*, 2011). The *tyf* gene was identified during an overexpression screen from a fly line producing long rhythms of locomotor activity. Further experiments restricting the overexpression of *tyf* to the PDF-positive LNvs only, recapitulated the long-period phenotype. Moreover, restoring wild-type *tyf* expression only in the PDF neurons rescued the *tyf*-null phenotype, suggesting that these are the cells where TYF is mainly required. Loss of TYF resulted in a dramatic reduction in PER (but not *per* mRNA) levels, especially in the LNvs, although PER cycling was dampened but not abolished. Immunoprecipitation experiments showed that TYF was associated directly with the PolyA Binding Protein (PABP) and indirectly with the 5′-cap-associating translation initiation factor eIF4G. Also *per* and to a lesser extent *tim* mRNAs co-immunoprecipitated together with TYF. TYF does not have any known RNA-binding motif; hence, its interaction with mRNAs is likely to be indirect. TYF was found, at least in part, cosedimenting with polysomes, and it was able to increase LUCIFERASE translation, both in cell culture and *in vitro*, when tethered to a luciferase reporter RNA. To explain the low level of PER in *tyf*-null mutants, it is hypothesized that *per* mRNA is bound by a translation inhibitor immediately after transcription. TYF would then recognize the inhibitor and displace it after binding to PABP, eIF4G, and other unknown factors, thus achieving efficient translation (Lim *et al.*, 2011).

F. Posttranslational regulation of clock proteins (part I): PER and TIM

There is growing agreement among chronobiologists that posttranslational modifications of clock proteins are the most important factors in maintaining circadian rhythmicity. For instance, behavioral rhythms and molecular oscillations of PER and TIM proteins do not require *per* or *tim* mRNA cycling (Frisch *et al.*, 1994; Yang and Sehgal, 2001). Conversely, when PER or TIM are overexpressed to the extent of abolishing protein cycling, arrhythmic behavior is triggered (Blanchardon *et al.*, 2001; Kaneko and Hall, 2000; Yang and Sehgal, 2001). Several kinases, phosphatases, and E3-ubiquitin ligases have been identified as controlling subcellular localization, stability, and ultimately activity of different clock components. This explains why PER and TIM cycling and rhythmic behavior may be maintained in spite of steady transcription, unless levels of those proteins become too high.

1. DOUBLE-TIME and NEMO

The first discovered and best characterized of these enzymes is the kinase DOUBLE-TIME (DBT), homologue of mammalian CASEIN KINASE 1ε (Kloss *et al.*, 1998; Price *et al.*, 1998). Hypomorphic mutations of *dbt* lengthen, shorten, or abrogate the period of locomotor activity depending on the specific molecular lesion (Kloss *et al.*, 1998; Price *et al.*, 1998; Rothenfluh *et al.*, 2000a). DBT physically interacts with PER (Kloss *et al.*, 2001) and phosphorylates it

(Ko *et al.*, 2002; Price *et al.*, 1998), slowing down PER accumulation in the cytoplasm by promoting its degradation (Ko *et al.*, 2002; Price *et al.*, 1998; Suri *et al.*, 2000). Considering that PER levels are constitutively low (Price *et al.*, 1995) and cytoplasmic (Saez and Young, 1996) in tim^0, and that TIM strongly binds to PER (Gekakis *et al.*, 1995; Myers *et al.*, 1995; Sehgal *et al.*, 1994), it has been proposed that rising TIM levels at the beginning of the night allow the formation of a complex that is resistant to the action of DBT and promotes nuclear accumulation (Gekakis *et al.*, 1995; Kloss *et al.*, 1998, 2001; Myers *et al.*, 1995; Price *et al.*, 1998; Sehgal *et al.*, 1994). Nuclear entry of PER and TIM is under considerable control, possibly because it gates the only known (but not necessarily exclusive) molecular function of the two proteins. It has been demonstrated that phosphorylation by DBT not only delays PER accumulation in the cytoplasm, but it is also sufficient *per se* to prevent nuclear entry. In fact, PER becomes nuclear independently of TIM in flies where DBT activity is strongly but not totally (in order to prevent excessive accumulation of PER) compromised (Cyran *et al.*, 2005). The timing of PER nuclear entry is also affected by CASEIN KINASE 2 (CK2) phosphorylation (Akten *et al.*, 2003; Lin *et al.*, 2002, 2005) as well as phosphorylation by an unknown (although recent developments suggest it might be NEMO) proline kinase that targets the Serine at position 661 (Ko *et al.*, 2010). PER is subject to multiple complex phosphorylation events that influence one another. Phosphorylation at Ser661 paves the way for a further phosphorylation event at Ser657 by another kinase, SHAGGY (SGG), the homologue of GLYCOGEN SYNTHASE KINASE-3β (GSK-3β) (Ko *et al.*, 2010). It is not clear whether PER and TIM enter the nucleus as a complex or independently as the experimental results are open to both interpretations. For instance, although TIM accumulates in the nuclear compartment in anticipation of PER (Meyer *et al.*, 2006), it also shuttles back to the cytoplasm (due to a nuclear export signal) until PER is securely nuclear (Ashmore *et al.*, 2003). However, DBT seems to keep hold of PER during the transfer, probably as part of a multiprotein complex (Kloss *et al.*, 2001), and continues its phosphorylation program in the nuclear compartment, ultimately leading to the complete degradation of PER (Bao *et al.*, 2001; Price *et al.*, 1998; Rothenfluh *et al.*, 2000a; Suri *et al.*, 2000) Yet, in the nucleus, PER is less dependent on TIM for stability than in the cytoplasm. After light-induced degradation of TIM at the beginning of the day (see below), highly phosphorylated PER persists for a few more hours into the nucleus, becoming, in the absence of TIM, a stronger repressor of the CLK–CYC activator (Rothenfluh *et al.*, 2000b). The key element driving DBT-mediated degradation of PER is the phosphorylation at Ser47, perhaps with the cooperation of a few adjacent phosphorylated sites, also targeted by DBT (Chiu *et al.*, 2008). When Ser47 is phosphorylated the F-box protein SLIMB, an E3-ubiquitin ligase can bind to PER and target it to the 26S proteosome for degradation (Chiu *et al.*, 2008; Grima *et al.*, 2002; Ko *et al.*, 2002). The fact that phosphorylation of one (or just a few) residue is sufficient for the binding of SLIMB is a dilemma, as a simple SLIMB-ON-SLIMB-OFF mechanism is difficult to reconcile with the observed complexity of PER phosphorylation and turnover. A partial answer to this, still mechanistically unclear, resides in a device that time-delays phosphorylation at Ser47 following phosphorylation at a cluster of distal DBT target sites. This cluster is centered around the classic *per*^{Short} (*per*^{S}) mutation encoding the substitution of Ser589 into Asn (Baylies *et al.*, 1987; Yu *et al.*, 1987a,b), which in flies is responsible for behavioral rhythms of about 19 h (Konopka and Benzer, 1971). Additional missense mutations in the same region also produce short-period phenotypes (Baylies *et al.*, 1992; Rothenfluh *et al.*, 2000a). Three sites, Thr583, Ser585, and Ser589, were shown to be phosphorylated by DBT in S2 cells (Chiu *et al.*, 2008) and constitute a discrete phosphocluster. Interestingly, when these sites were mutagenized into Ala *in vitro* to prevent phosphorylation, each caused faster phosphorylation at Ser47 as assessed by faster binding of SLIMB to PER in S2 cells. Another substitution, Ser596 to Ala, was responsible for an even larger effect (Chiu *et al.*, 2011). Using phosphospecific antibodies, it was possible to demonstrate that phosphorylation at Ser596 enhanced DBT-dependent phosphorylation at Ser589 in S2 cells, whereas blocking phosphorylation at Ser596 strongly delayed phosphorylation at Ser589 in S2 cells and abolished it altogether in flies. Moreover, in flies collected at different times during the daily cycle, it was shown that phosphorylation at Ser596 precedes that at Ser589, which in turn is seen before phosphorylation at Ser47. As expected, phosphorylation of Ser47 could be detected 4–8 h earlier in flies carrying a Ser to Ala mutation in position 596 or 589 compared to those carrying wild-type PER. Serine at position 596 is followed by a proline residue, indicating that Ser596 is phosphorylated by a "pro-directed" kinase. There are 34 known/predicted such kinases in *Drosophila*. An RNAi screen was performed to identify the relevant kinase, under the assumption that the knockdown would phenocopy the Ser596 to Ala mutation resulting in a shorter behavioral period. Only RNAi directed against the kinase NMO produced a shorter period and reduced phosphorylation at Ser596. From these data, a model emerges where phosphorylation at Ser596 by NMO promotes phosphorylation by DBT at the "*per*^{S} domain" cluster, namely at Ser589, at Ser585 and perhaps at Thr583. This phosphocluster somehow delays DBT-mediated phosphorylation at other sites including Ser47, thus delaying PER-SLIMB interaction and PER degradation (Chiu *et al.*, 2011) (Fig. 4.6).

2. CASEIN KINASE 2 and SHAGGY

Other kinases are involved in the posttranslational regulation of PER but also have an effect on TIM and *vice versa*, although the joint regulation of the two clock proteins makes it difficult to separate direct and indirect effects. CK2 is a tetramer

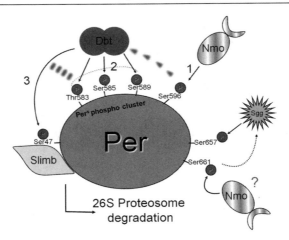

Figure 4.6. Important phosphorylation sites on PER. Every circadian kinase can phosphorylate PER. However, not all phosphorylation sites have been mapped or their functional significance identified. Here, we show the main phosphorylation sites having a known functional effect on PER. Phosphorylation at Ser661 by a proline-directed kinase (perhaps NMO, hence the question mark) prime the phosphorylation of Ser657 by SGG, promoting nuclear translocation. Phosphorylation at Ser596 by NMO promotes (1 and sequence of red arrowheds) phosphorylation by DBT of the PER^S phosphocluster (Thr583, Ser585, and Ser589). In turn, this delays (2 and sequence of red rectangles) phosphorylation of Ser47 also by DBT (3). The latter is the key phosphorylation event leading to SLIMB binding and proteasome mediated degradation of PER. (See Color Insert.)

composed of two α-catalytic subunits and two β-regulatory subunits. Hypomorphic mutations affecting the α (*Tik*) or the β (*Andante*) subunits produced a long period of locomotor activity, reduced PER and TIM degradation and delayed nuclear entry (Akten *et al.*, 2003; Lin *et al.*, 2002, 2005; Meissner *et al.*, 2008; Smith *et al.*, 2008). Although *in vitro* phosphorylation experiments suggested that PER is a direct target of CK2 (Lin *et al.*, 2002), reduction of CK2 activity *in vivo* led to increased TIM levels in the absence of PER (*per⁰* flies), but PER levels were not increased in the absence of TIM (*tim⁰* flies) (Meissner *et al.*, 2008).

TIM is the major circadian target for the *Drosophila* GSK-3β homologue SGG (Martinek *et al.*, 2001). Flies overexpressing SGG show a shorter period of locomotor activity due to increased nuclear translocation of TIM and PER. Conversely, reduction of SGG activity results in a longer period. SGG has been shown to phosphorylate TIM *in vitro* (Martinek *et al.*, 2001). *In vivo*, hypomorphic mutations of *sgg* reduce TIM phosphorylation and increase levels of both TIM and PER in DD, suggesting that SGG likely contributes to TIM turnover in the dark (Martinek *et al.*, 2001).

3. Protein phosphatases

Where there is a kinase, there always is a phosphatase and indeed, phosphatases are involved in the regulation of clock proteins. Protein Phosphatase 2A (PP2A) is a heterotrimeric enzyme composed of a highly conserved catalytic subunit, a variable regulatory subunit, and a structural subunit. Its involvement in the clock was hypothesized because PP2A is part of the Wnt signaling pathway, like DBT and SGG (Sathyanarayanan et al., 2004). Overexpression of *twins* (*tws*) encoding for one of the regulatory subunits of PP2A in the PDF-positive LNvs, caused a short period of locomotor activity that degenerated into arrhythmia after 4–7 days in constant darkness. Similar results were obtained by overexpressing the catalytic subunit MUTAGENIC STAR (MTS) that molecularly resulted in stable and nuclear PER. Conversely, decreasing PP2A activity by overexpression of a dominant-negative version of MTS (MTS^{DN}), resulted in a long period of locomotor activity and in lower levels of PER. PER protein expressed in bacteria and phosphorylated *in vitro* by the active form of mammalian CKIδ was dephosphorylated when incubated with PP2A, suggesting that the effect of PP2A on PER is likely to be direct. Interestingly, PP2A activity might be circadian as both *tws* and *widerborst* (*wdb*, encoding for another regulatory subunit) are rhythmically expressed, possibly under CLK–CYC control (Sathyanarayanan et al., 2004).

The other major Eukaryote phosphatase, Protein Phosphatase 1 (PP1), has also been shown to be implicated in the circadian clock. PP1 is an oligomeric complex comprising a core enzyme, the catalytic subunit that can bind to a spectrum of regulatory subunits. In *Drosophila*, four genes encode a catalytic subunit of PP1 (PP1c) and are named according to their chromosomal loci: 9C (also called *flapwing*), 13C, 87B, and 96A. The four PP1c isoforms can compensate for each other and only by affecting the whole catalytic activity of PP1 do phenotypes emerge. An endogenous nuclear inhibitor of PP1 (NIPP1), is a strong and specific inhibitor of all PP1c isoforms; therefore, its overexpression was used to inhibit PP1 function. Reduced PP1 activity resulted in decreased TIM levels and in a long ($\sim$26.2 h) and weak activity period. Also, the accumulation of nuclear TIM was delayed, although the onset of nuclear entry remained unchanged. Co-immunoprecipitation experiments in S2R^{+} (a variant of S2) cells showed that TIM can physically interact with any of the four PP1c isoforms. Moreover, co-immunoprecipitated TIM that was then phosphorylated with mammalian CKIδ and GSK3β *in vitro* was dephosphorylated by the addition of purified PP1. Therefore, inhibition of PP1 should result in increased levels of phosphorylated TIM akin to the overexpression of SGG. However, as discussed above, the two manipulations caused different effects on the nuclear accumulation of TIM (faster after SGG overexpression and delayed after PP1 inhibition) and opposite outcomes on the circadian period (shortening of period after SGG overexpression and lengthening after PP1 inhibition). A possible interpretation could be that, in contrast to the *in vitro* results, PP1 does not dephosphorylate SGG target sites on TIM *in vivo*. Yet, flies coexpressing NIPP1 and SGG exhibited an average period of 19.5 h which is only ∼1.3 h longer than that of flies overexpressing SGG alone. Thus, the period-lengthening effect of NIPP1 was reduced by ∼50% in a SGG-overexpressing background. In addition, overexpression of NIPP1 did not reduce rhythmicity in this background to the same extent as in the wild-type background. These results highlight a genetic interaction between PP1 and SGG activity, but the molecular details behind these interactions are not currently understood (Fang *et al.*, 2007).

4. SLIMB and CTRIP

We have already mentioned that SLIMB (SUPERNUMERARY LIMBS), an F-box/WD40-repeat protein, binds to PER after phosphorylation of Ser47 by DBT. SLIMB function as part of a SCF complex, a multiprotein E3-ubiquitin ligase, whose function is to attach ubiquitin residues (usually, sequentially, forming a chain) to lysine in a target protein. In the complex, the E2-ubiquitin conjugating enzyme catalyzes the reaction, while the F-box protein provides target specificity (Skowyra *et al.*, 1997). *slimb* is an essential gene but *slimb*-null mutants can progress through development if SLIMB is provided *via* a heat-shock construct activated by 37 °C pulses. The adult flies can then be tested without heat shock. Lack of SLIMB (Grima *et al.*, 2002) or downregulation *via* expression of a dominant-negative version of the protein (Ko *et al.*, 2002), resulted in arrhythmic behavior. Western blot analyses showed that highly phosphorylated PER and TIM isoforms were present constitutively in *slimb*-null mutants in DD but not in LD, suggesting that SLIMB is involved in the turnover of hyperphosphorylated forms of both proteins but limited to DD conditions (Grima *et al.*, 2002). Co-immunoprecipitation experiments carried out *in vivo* showed that SLIMB can bind to hyperphosphorylated forms of PER (Grima *et al.*, 2002). In S2 cells, PER can be efficiently phosphorylated by DBT (Ko *et al.*, 2002). Co-immunoprecipitation in this cellular system confirmed the results above (Ko *et al.*, 2002) and extended them by showing that phosphorylation at Ser47 is a key step in regulating SLIMB binding to PER, although DBT-phosphorylated sites within the first 100 amino acids also contribute to the binding (Chiu *et al.*, 2008).

CTRIP is another E3 ligase regulating the turnover of PER; additionally, it is also involved in the degradation of CLOCK. The discovery of *ctrip* stemmed from the identification of the locus affected by P-element insertion in the enhancer trap lines *gal1118* and *gal1501*. Both have been described as mainly driving expression of reporter genes in the PDF-positive LNvs and, more weakly, in a few additional clock and nonclock cells (Blanchardon *et al.*, 2001; Lamaze *et al.*, 2011). In both lines, the target of insertion is the gene *CG42574*, which shows sequence homology to human *trip12*. The latter encodes for a E3-ubiquitin ligase that contains Armadillo repeats, a WWE protein–protein interaction domain and a carboxy-terminal HECT domain; CG42574 was then renamed *ctrip* (Lamaze *et al.*, 2011). Loss of *ctrip* is lethal at the pupal stage, but the analysis of clock neurons in *ctrip*-null third-instar larvae showed delayed cycling and higher levels of PER and CLK limited to DD. Experiments conducted by RNAi in adults confirmed the observations in larvae, in addition, an excess of hyper-phosphorylated TIM was found. Transcripts levels of the evening genes were augmented but not seriously delayed in *ctrip* RNAi flies, suggesting that the parallel increase of the activator, CLK, and of the repressor, PER, had buffered each other, and that increases in PER and CLK levels were probably posttrans-lational. Downregulating *ctrip* by RNAi in per^0 flies showed that the effects on TIM were PER dependent but increased levels of CLK still persisted in this background (Lamaze *et al.*, 2011).

G. Posttranslational regulation of clock proteins (part II): CLK

Before the discovery of CLK and CYC, the description of the PAS domain (a structure promoting dimerization), of which PER is a founding member, and the realization that TIM did not harbor such a region, suggested an attractive mechanism to explain the negative feedback loop. The hypothesis was that the putative transcription factor driving the expression of *per* and *tim* would carry a PAS domain and that PER, either alone or bound to TIM, would function by sequestering it (Huang *et al.*, 1993). The discovery that both CLK (Allada *et al.*, 1998) and CYC (Rutila *et al.*, 1998) had a PAS domain, seemed in line with the expectations, but this model was short lived. CLK was shown to cycle in LD and DD, with a peak of abundance at the beginning of the (actual or subjective) day and a trough at the beginning of the (actual or subjective) night, this is 3–4 h later than PER. Moreover, the increase in levels paralleled the increase in electrophoretic mobility, corresponding to the accumulation of highly phosphor-ylated forms of the protein (Lee *et al.*, 1998). On the contrary, CYC was found to be expressed at constitutively high levels and without changes in its phosphor-ylation status (Bae *et al.*, 2000). Co-immunoprecipitation experiments also showed that CLK and CYC were bound together at all times, that PER–TIM interactions with CLK increased at the end of the night, and that CLK was the main interacting partner of PER–TIM (Bae *et al.*, 2000). Further experiments, including chromatin immunoprecipitation, showed that the amount of CLK was stable across the 24 h, and that the apparent cycle in abundance was an artifact created by buffers that were too mild to extract CLK efficiently when strongly bound to the chromatin (Yu *et al.*, 2006). Therefore, although there is not a cycle in protein abundance, there is a cycle in the interaction of CLK–CYC with the DNA. It has been shown that when PER–TIM bind to CLK–CYC, the complex acquires very quickly a repressed state although it remains attached to the E-boxes for a few more hours, before being released from the DNA (Menet et al., 2010). However, the experiments detailed below show that inhibition of transcription is not dependent upon the physical association of PER–TIM with CLK–CYC, but it requires the phosphorylation of CLK (Yu et al., 2009). The main drive in the interaction between PER–TIM and CLK–CYC is the physical association between PER and CLK which does not involve the PAS domain but the C-terminus of PER. Using deletion experiments coupled with a transcription assay in S2 cells, it was first determined that a region of PER encompassing amino acids 764–1034 (called the CLK–CYC Inhibition Domain or CCID) is necessary and sufficient for the inhibition of CLK–CYC-mediated transcription (Chang and Reppert, 2003). This region was better annotated by the finding that the stretch between amino acids 926–977 (called CLK Binding Domain or CBD) is sufficient for efficient binding to CLK and transcription inhibition (Sun et al., 2010). Moreover, the physical interaction between PER and DBT requires the region between amino acids 755–809 (PER–DBT Binding Domain, aka PDBD), corresponding to the beginning of the CCID (Kim et al., 2007). Transgenic flies expressing a form of PER unable to bind to DBT (PER-ΔPDBD) were arrhythmic and unable to repress CLK–CYC-mediated transcription, even though PER-ΔPDBD was still able to co-immunoprecipitate CLK and TIM (Kim et al., 2007). On the contrary, flies expressing PER-ΔCBD, a form of PER unable to bind to CLK, were rhythmic, albeit with longer period, and showed quasi-normal PER/TIM biochemical rhythms (Sun et al., 2010). TIM had been shown to bind CLK independently of PER (Lee et al., 1998); thus, TIM could function as a bridge between CLK and PER-ΔCBD, allowing DBT to come into the complex. In support of this hypothesis, PER-ΔCBD-mediated repression of CLK–CYC was rapidly attenuated by light, which is expected considering TIM sensitivity to light (see below) (Sun et al., 2010). In flies expressing PER-ΔPDBD, PER, and CLK were hypophosphorylated, and CLK–CYC-mediated transcription continued unabated. Thus, DBT is an essential component for transcriptional repression. In a following experiment, DBT catalytic activity was removed from clock cells by expressing (only in those cells to avoid lethality) DBT$^{K/R}$, a dominant-negative form of DBT, in mutants carrying a strong dbt-hypomorph allele. In these flies, PER was still hypophosphorylated, whereas CLK resulted hyperphosphorylated, and its transcriptional activity became repressed (Yu et al., 2009). Because DBT$^{K/R}$ was not functional, one or more additional kinases must have been responsible for the extensive phosphorylation and the consequent inactivation of CLK. However, although DBT$^{K/R}$ lacked catalytic function, it maintained binding specificity, suggesting that the latter was required by the activity of the additional kinases, possibly as a bridge to reach CLK (Yu et al., 2009). An RNAi screening, corroborated by co-immunoprecipitation and Western blot analyses of clock proteins, identified NMO as one of these new kinases.

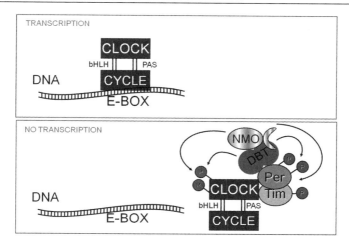

Figure 4.7. CLK–CYC-mediated transcription is regulated by phosphorylation of CLK. The current model suggests that CLK–CYC-mediated transcription occurs early at night when CLK is hypophosphorylated. Late at night, a complex composed of PER, TIM, DBT, and NMO promotes the phosphorylation of CLK. Hyperphosphorylation of CLK correlates with the arrest of transcription activity, the removal of the CLK–CYC complex from E-boxes and the degradation of CLK. See text for further details. (See Color Insert.)

For instance, misexpression of NMO affected the phosphorylation state of PER, TIM, and CLK. Overexpressed NMO co-immunoprecipitated with PER, TIM, and CLK at times when CLK–CYC-dependent transcription is repressed. Down-regulation of NMO *via* RNAi caused higher abundance of CLK, whereas its overexpression resulted in lower CLK levels (Yu *et al.*, 2011). This suggests that NMO is important for CLK stability. However, both overexpression and down-regulation of NMO increased the phosphorylation of CLK (Yu *et al.*, 2011), suggesting that NMO might have a complex effect on CLK phosphorylation. Currently, there is no direct evidence that NMO phosphorylation inhibits CLK–CYC transcription (Fig. 4.7).

III. ENTRAINMENT

The molecular mechanisms described so far occur in LD, but they largely continue also under DD (in jargon, these are called free-run conditions because not only light but also temperature are maintained constant), explaining a fundamental property for biological clocks, self-sustainability. Free-run conditions do not exist in nature for *D. melanogaster*; thus, the intrinsic rhythmicity of the clock perhaps arose from the need to maintain high-amplitude cycles even at times of reduced strength in the environmental light and temperature signals (think about a dull, rainy day in autumn). Another major function of the clock is to harmonize the internal milieu with the external environment. Thus, biological clocks have the ability to respond appropriately to even small changes of the photo- and thermoperiod. Not only, once "trained" to a new regime, clocks are able to "remember" and anticipate the daily cycles in light and/or temperature. This is more than synchronization, we call this complex property entrainment.

A. Anatomy of the clock

Before proceeding further, we need to introduce in some detail the anatomy of the clock. Note that although we generically refer to it as "the clock," in reality, our discussion is focused on the system maintaining circadian rhythms of locomotor activity in the adult fly. This is just one of the many rhythmic behaviors and physiological processes that possibly rely on different types of cells and molecular components.

     Previous to the cloning of *per*, heroic transplantation experiments (Handler and Konopka, 1979) and genetics mosaic analyses (Konopka *et al.*, 1983) identified the head as the center responsible for locomotor activity. After the cloning of *per* and the production of antibodies, immunohistochemical investigations showed that PER is widely distributed in the head, being present in photoreceptor cells of the eyes and ocelli, in glia, and in neurons (Ewer *et al.*, 1992; Siwicki *et al.*, 1988). Experiments with mutants, however, ruled out photoreceptor cells as necessary for rhythmicity and, curiously, for entrainment (Dushay *et al.*, 1989; Helfrich, 1986; Helfrich and Engelmann, 1983; Vosshall and Young, 1995; Wheeler *et al.*, 1993; Yang *et al.*, 1998). Glia was also not considered the main determinant of behavioral rhythms; an analysis of flies that were random per^0/per^+ mosaics showed that robust rhythmicity was always associated with per^+ tissue in a lateral region in between the central brain and the optic lobe and flies with per^+ immunoreactivity limited to glia cells, showed only weak rhythmicity (Ewer *et al.*, 1992). As further evidence, a promoter-less *per* transgene that was able, in per^0 flies, to rescue rhythmic PER expression in lateral neurons (LNs) but not in glia, was also able to produce rhythmic behavior (Frisch *et al.*, 1994). This suggested that the PER-positive neurons were the focus of the behavioral clock. After the identification and cloning of TIM (Myers *et al.*, 1995; Sehgal *et al.*, 1994), immunohistochemical experiments showed that TIM immunoreactivity largely overlapped with the one of PER (Hunter-Ensor *et al.*, 1996; Kaneko and Hall, 2000; Kaneko *et al.*, 1997; Myers *et al.*, 1996; Yang *et al.*, 1998). The PER/TIM immunopositive cells were then crowned as the clock neurons. Furthermore, a landmark experiment showing that *disconnected* (a mutation causing the loss of many neurons in the brain, including lateral

PER-positive neurons) flies are rhythmic if at least one of the LNs is spared, pointed to the LNs (the s-LNvs in particular) as the most important of all clock cells (Helfrich-Forster, 1998). The observations cited above collectively provide convincing evidence that the canonical clock neurons are major determinants of rhythmicity. However, they do not refute the involvement of other cells. For instance, the importance of glia for circadian function has recently been vindicated (Ng *et al.*, 2011; Suh and Jackson, 2007). Moreover, some scattered evidence has accumulated over the years, suggesting that more neurons could be part of the circadian neuronal network. There are examples where some uncharacterized neurons have been manipulated in addition to some canonical clock neurons and seem to have contributed to the rescue or suppression of rhythmicity (Fujii and Amrein, 2010; Shang *et al.*, 2008; Taghert *et al.*, 2001; Vosshall and Young, 1995). Moreover, to overexpress or downregulate genes in clock cells (see above), we often rely on "drivers" based upon the *tim* promoter. The (strongly) immunoreactive TIM cells are a fraction compared to those where the *tim* promoter is active (Kaneko and Hall, 2000); therefore, the results of many published experiments are based on the manipulation of a larger number of cells than the canonical clock neurons.

The *bona fide* clock neurons are divided in three groups of LNs and three groups of dorsal neurons (DNs) (Ewer *et al.*, 1992; Frisch *et al.*, 1994; Siwicki *et al.*, 1988; Zerr *et al.*, 1990). The LNs are subdivided in dorsal (LNds, 1 cluster of ~6 cells) and ventral and the latter also in large (l-LNvs, 4–5 cells) and small (s-LNvs, 4 cells). One additional small neuron is usually found very close to the l-LNvs but, unlike the other s-LNvs, does not express the neuropeptide PDF (Kaneko and Hall, 2000); we refer to it as the PDF-null LNv (pn-LNv). The DNs, which are more dorsal but also more posterior in the brain, are subdivided in DN1s (~16 cells), DN2s (2 cells), and DN3s (~40 cells) (Helfrich-Forster, 2003, 2005; Kaneko and Hall, 2000). Three lateral posterior neurons (LPNs) constitute a further cluster which expresses high levels of TIM (Kaneko and Hall, 2000) and only low levels of PER (Helfrich-Forster, 2005; Shafer *et al.*, 2006). In spite of their anatomical proximity, cells within a cluster might not be functionally identical as they might be differentiated by the type of neurotransmitters/neuromodulators they release, the complement of genes they express, the stability of proteins they produce, and the projections of their dendrites and axons (see Hermann *et al.*, 2012; Peschel and Helfrich-Forster, 2011) (Fig. 4.8)

B. Entrainment to light

The discovery that TIM, a fundamental component of the TTL, is degraded after exposure to light prompted a simple cell-autonomous model to explain photic entrainment (Hunter-Ensor *et al.*, 1996; Marrus *et al.*, 1996; Myers *et al.*, 1996; Yang *et al.*, 1998). At the beginning of the day, exposure to light degrades TIM,

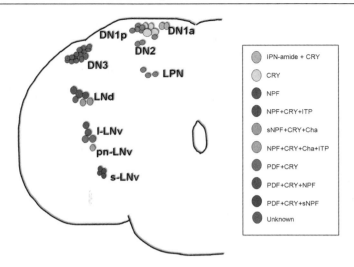

Figure 4.8. Circadian neurons, proteins, and neuropeptides in the brain of *D. melanogaster*. A simplified model of the circadian neuronal clusters as described in the text. Note that neurons from the same cluster express different proteins/neuropeptides. IPN-amide; CRY, CRYPTOCHROME; NPF, NEUROPEPTIDE F; ITP, ION TRANSPORT PEPTIDE; Cha, CHOLINE ACETYLTRANSFERASE; sNPF, SHORT NEUROPEPTIDE F. Cells with unknown peptidergic content are shown in gray. Drawn from Peschel and Helfrich-Forster (2011) and Hermann *et al.* (2012). (See Color Insert.)

which allows DBT-mediated degradation of PER to take place. Thus, a new day brings the inhibition of CLK–CYC-dependent transcription to a halt and the start of a new cycle. This simple process can explain the synchronizing effect of light, because when the lights are turned on, irrespective of the stage of progression of the endogenous cycle, TIM and (indirectly) PER plunge to their lowest level, signaling to the clock that a new day is starting, with CLK–CYC-mediated transcription following suite. Moreover, under constant light conditions, the extensive degradation of TIM and PER would be responsible for longer behavioral periods (as PER–TIM would require more time to accumulate to sufficient levels to trigger CLK–CYC inhibition) or arrhythmicity (when PER–TIM levels become negligible), depending on the intensity of the light. The same principle also informs another property of light entrainment, the phase shift response. This is a change in the phase of the endogenous rhythm (thus requiring DD to become evident) after delivering a light pulse in the real or subjective night (the time corresponding to night but in DD). For simplicity, let's refer to the conditions of a so-called anchored protocol. Flies are entrained under LD before being released into DD. During the last night, a short light pulse (usually 5–20 min) is delivered, and the DD activity is recorded in the following days. If the light pulse is delivered during the early night, it will give rise to a phase delay, namely the flies will wait several hours before starting their activity on the following day. If the pulse is delivered at the end of the night, the flies will anticipate the onset of activity showing a phase advance. In the middle of the night, there is a dead zone when the pulse does not elicit any shift, the dead zone extends to the day or, under constant conditions, to most of the subjective day (Young, 1998). Western blot analyses have shown that corresponding phase delays and advances also occur in the molecular rhythm of TIM (Myers *et al.*, 1996). When a pulse is delivered at the beginning of the night, there are large reserves of *tim* and *per* RNAs, but only small amounts of proteins are available. The light pulse destroys TIM, hence PER, pushing the clock backwards. The time it takes for the levels of TIM and PER to rise back determines the shift. Moreover, TIM and PER are cytoplasmic and hypophosphorylated early at night; therefore, not only protein levels but also posttranslational modifications will contribute to the length of the delay, which can reach a maximum of $\sim$4 h. Late at night, TIM and PER are abundant, nuclear, and hyperphosphorylated, whereas mRNA levels are low. After the light pulse, TIM and PER disappear quickly, but they cannot be replenished by their exhausted mRNA reserves. The clock is pushed forward, resulting in an advance that can reach a maximum of $\sim$2 h (Wijnen and Young, 2008).

      Considerable efforts have been directed towards describing the molecular details of this model. It was established that phosphorylation by a tyrosine kinase (never identified) and the proteosome are required for light-driven degradation of TIM (Naidoo *et al.*, 1999). However, the major step forward has been the discovery of cryptochrome in *Drosophila*, a blue-light sensitive protein related to 6-4 photolyases, already known as an important photopigment in plants (Ahmad and Cashmore, 1996). During a mutagenesis screening to identify genes controlling the cycling expression of circadian reporters, a mutant was discovered where PER and TIM were largely prevented from degradation both under LD and DD (Stanewsky *et al.*, 1998). The mutant was identified as a strong hypomorph of the *cryptochrome* (*cry*) gene and was called *cry*^{baby} or *cry*^{b} for short (Stanewsky *et al.*, 1998). The rhythmic expression of *cry* mRNA with a peak at the end of the night, which was disrupted in clock mutants, disclosed it as an evening gene (Emery *et al.*, 1998). On the contrary, the cycling of CRY was shown to be autonomous from the clock but dependent on light, as protein levels raise at night but decline during the day (Emery *et al.*, 1998). In *cry*^{b} flies, Western blot analyses showed a combination of hypo- and hyperphosphorylated forms of PER and TIM and either no cycling (Stanewsky *et al.*, 1998) or reduced cycling (Zhu *et al.*, 2008) for both proteins under LD and DD conditions. Interestingly, PER and TIM oscillations could be rescued in *cry*^{b} flies by temperature cycles (in darkness) and persisted for at least 1 day when flies were switched to constant temperature. Lack of PER and TIM cycling in LD and DD should result in arrhythmic behavior (or in a long period in those genetic backgrounds where cycling is reduced but not abolished), but cry^b flies are perfectly rhythmic within the wild-type range. This was explained by the discovery that the main clock neurons show significant, albeit reduced, cycling of the two clock proteins (Helfrich-Forster et al., 2001; Stanewsky et al., 1998). However, cry^b flies display some significant behavioral defect, being rhythmic (rather than arrhythmic) under constant high levels of light (Emery et al., 2000a,b) and irresponsive (no phase shift) to pulses of light (Stanewsky et al., 1998). Nevertheless, they entrain normally to LD cycles, unless they are combined with mutations that damage the functioning of the external photoreceptors (Helfrich-Forster et al., 2001; Stanewsky et al., 1998). These observations, together with the evidence that light-activated CRY binds to TIM (Ceriani et al., 1999) and that this binding is important for light-dependent proteosomal degradation of TIM (Busza et al., 2004), established CRY as the "dedicated circadian photoreceptor" (Emery et al., 2000b) in flies. Translated from jargon, this meant that the activation of CRY by light was considered the key step of the light-dependent and cell-autonomous degradation of TIM. The goal at this point was to identify the molecular partners involved in transmitting a degradation signal from light-activated CRY to TIM. Another mutagenesis screening identified one such component, which was named JETLAG (JET) (Koh et al., 2006). JET is an F-box protein with leucine-rich repeats, and a putative component of a Skp1/Cullin/F-box (SCF) E3 ubiquitin ligase complex. Two mutations of the jet locus, one more common (jet^c) the other rarer (jet^r), resulted in flies showing rhythmic behavior under constant light, reduced phase shifts in response to light pulses, and reduced light-dependent degradation of TIM (Koh et al., 2006). An interesting twist of this story is that the behavioral effects of the jet mutations become visible only in the presence of a particular variant of TIM (Peschel et al., 2006). An insertion/deletion polymorphism found in the tim locus of natural populations can generate two forms of the TIM protein that differ at the N-terminus for the presence (the long form) or absence (the short form) of 23 amino acids (Rosato et al., 1997; Sandrelli et al., 2007; Tauber et al., 2007). The ancestral allele, s-tim, encodes for the short form. An insertion of a G nucleotide, a few residues upstream of the ATG translational start for the short form, allows another ATG, 23 codons upstream, to become in frame. This evolutionary new allele, sl-tim, encodes and can produce both S- and L-TIM isoforms. The latter (L-TIM) was found to be less light sensitive according to different parameters and its physical interaction with CRY was weaker (Peschel et al., 2006; Sandrelli et al., 2007; Tauber et al., 2007). Going back to the jet alleles, these produce a phenotype only in combination with sl-tim (Peschel et al., 2006). To explain these observations and others, Peischel and coworkers proposed the following model. Light-activated CRY (CRY*) binds to TIM inducing a posttranslational modification (TIM*) that is a prerequisite for JET:TIM* association. JET can dimerize either with TIM* or CRY* targeting its partner for proteosomal degradation. JET has higher affinity for TIM* than for CRY*, but formation of TIM* requires prior interaction with CRY*. S-TIM binds strongly to CRY*, thus it is effectively activated, incorporated into the JET:S-TIM*complex, and targeted for degradation. CRY* is degraded after S-TIM, explaining why CRY is more stable in a *s-tim* background. L-TIM binds more weakly to CRY*, thus there is less L-TIM* to compete with CRY* for JET binding. As JET:CRY* complexes are formed and less CRY* becomes available, formation of L-TIM* is further reduced, which explains the increased stability of L-TIM and the reduced light responses of *ls-tim* compared to *s-tim* flies (Peschel *et al.*, 2009) (Fig. 4.9).

The results discussed above seem to offer a molecular justification to the hypothesis that cell-autonomous, light-dependent degradation of TIM is the cause of light entrainment. However, there are a few observations that shake this widely accepted idea. First of all, inconsistencies emerge by considering the behavior of *cry*[b] flies. These flies are rhythmic under DD, entrain normally under LD but fail to phase shift after light pulses and do not become arrhythmic under constant light. If CRY is essential for light-driven degradation of TIM, how can these flies entrain to LD cycles? It has been suggested that external photoreceptors and other unknown photopigments that are possibly expressed directly in some of the circadian neurons could compensate for the lack of CRY allowing entrainment under LD conditions (Helfrich-Forster *et al.*, 2001), after all PER and TIM cycle in the clock neurons of *cry*[b] mutants (Helfrich-Forster *et al.*, 2001; Stanewsky *et al.*, 1998). The problem then becomes to explain why these same mechanisms fail under LL or after a light pulse. Moreover, when wild-type flies are released into LL, they show signs of molecular and behavioral rhythmicity for the first couple of days (Marrus *et al.*, 1996), which is incompatible with a model where light inexorably triggers TIM degradation and immediately stops the TTL. Equally, temperature cycles rescue behavioral rhythmicity

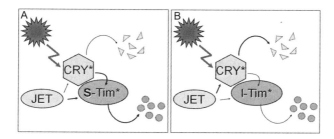

Figure 4.9. Model of light-mediated degradation of TIM and CRY. (A) The stronger affinity of S-TIM for light-activated CRY (CRY*) results in a stronger JET/S-TIM* interaction and faster light-mediated degradation of S-TIM*. (B) The weaker affinity of L-TIM for CRY* results in a stronger JET/CRY* interaction and faster light-mediated degradation of CRY* (Peschel *et al.*, 2009). (See Color Insert.)

and TIM cycling under LL (Yoshii *et al.*, 2002, 2005, 2007). Finally, anti-CRY immunolabeling experiments have revealed that many clock cells (three LNDs, five to six DN1s, most DN3s, both DN2s, and all the LPNs) show little or no expression of CRY (Benito *et al.*, 2008; Yoshii *et al.*, 2008), posing the question of how light-dependent degradation of TIM might proceed in those cells. One possibility is that light entrainment and TIM degradation primarily depend upon intercellular communications, and less on cell-autonomous events. In support of this view, it has been shown that light-activated CRY directly increases neuronal firing (Fogle *et al.*, 2011), which implicates CRY in cellular cross-talk, with a mechanism that is independent from the TTL. Other findings further suggest that light entrainment probably relies on network properties rather than on cell-autonomous mechanisms. It has been observed that TIM is not degraded at all in the s-LNvs after a light pulse delivered early in the night, whereas TIM disappears from many, albeit not all, clock cells from other groups (Tang *et al.*, 2010). This shows that TIM degradation in the s-LNvs is not necessary for behavioral phase delays. TIM degradation in s-LNvs is also not sufficient to generate phase delays since forcing the degradation of TIM by overexpressing JET in the s-LNvs of *jet^c* flies did not rescue the reduced phase shift defect of the mutant (Tang *et al.*, 2010). However, reducing the level of CRY by RNAi only in the s-LNvs showed that both CRY and the s-LNvs are necessary for phase delays (Tang *et al.*, 2010). Perhaps, limited degradation of TIM only occurs in the early evening, when the levels of CRY are low (CRY is light sensitive and accumulates during the night), and TIM is hypophosphorylated. As a test a light pulse was delivered at the end of the night and then the immunopositive TIM cells were counted. This resulted in about 50% of the s-LNvs showing no TIM. However, the other groups of neurons showed more TIM immunoreactive cells after the late night than after the early night light pulse (Tang *et al.*, 2010), in spite of the higher levels of CRY (Emery *et al.*, 1998) and the hyperphosphorylated TIM (Tang *et al.*, 2010). Overall, these results show that light entrainment is mainly a "system" response and undermine the hypothesis that TIM degradation is causal to entrainment, at least for the locomotor activity clock. Although not much information is available, it is possible that peripheral clocks rely more heavily on cell-autonomous TIM degradations to synchronize to light.

Further evidence that light entrainment is a property of the whole network rather than a cell-autonomous response comes from the discovery of *quasimodo* (*qsm*). This morphologically "hunchback" mutant (hence the name) was identified because of anomalous behavioral rhythmicity and persistent PER/TIM oscillations in the DNs and the eyes under constant light conditions (Chen *et al.*, 2011). QSM is a zona pellucida type of protein likely attached to the extracellular side of the cell membrane through a glycosyl phosphatidylinositol membrane anchor. It has been shown that the transcription of *qsm* is under clock control, but the regulation of the protein is seemingly mediated by light, as QSM levels increase immediately after lights on. However, the possibility exists that this apparent increase is just an artifact due to a conformation change picked up by the antibodies currently available (Chen *et al.*, 2011). QSM was detected in photoreceptor cells, in one to two LNds and in many DNs, but it was not present in the LNvs with the only exception of the pn-LNv. Since only some of the neurons expressing QSM also expressed CRY, this suggested that QSM might function independently of CRY. In fact, the light-induced increase of QSM and the significant reduction of TIM after QSM overexpression remained intact in *cry*-null flies. There are no clues on the molecular function of QSM. However, considering its predicted membrane localization, it is possible that a conformational change, triggered by a light-activated, membrane-bound photopigment, would allow QSM to interact with other membrane proteins, for example, ion channels, altering the electrical properties of the cell (Chen *et al.*, 2011). Therefore, as for CRY (although independently from it), the main effects of QSM, such as light entrainment and TIM degradation, would result from cell-to-cell communication.

C. Entrainment to temperature

Temperature cycles are able to entrain rhythmic locomotor activity behavior under constant darkness (Wheeler *et al.*, 1993) or constant light (Yoshii *et al.*, 2002, 2005, 2007), suggesting that temperature is a potent entraining stimulus. To investigate how temperature can entrain the clock, Sidote and colleagues delivered heat pulses either at the beginning or the end of the night and observed, at both times, a rapid decrease in the level of PER and TIM (Sidote *et al.*, 1998). However, the flies responded behaviorally, with a phase delay, only to early night pulses but did not advance their activity rhythm after a late-night heat shock. Thus, unlike light, temperature does not provoke a specific molecular change that can be univocally correlated with a behavioral response. As a result, there has not been a clear hypothesis about the mechanism by which temperature entrains the clock. In an attempt to identify the cells mediating temperature entrainment, Miyasako and colleagues adopted an interesting paradigm (Miyasako *et al.*, 2007). They exposed flies to both LD and temperature cycles (25 °C/20 °C), either in phase (20 °C at night, 25 °C during the day) or 6 h out of phase, with the thermophase (25 °C) preceding the switching on of lights. They then assayed the cycling of clock neurons using TIM immunoreactivity as a readout. What they observed was that the LNs generally stayed synchronous to the LD cycles in both conditions showing a peak of TIM immunoreactivity towards the end of the dark phase. Conversely, the DNs (the DN2s in particular) and the LPNs, phase advanced when challenged with nonsynchronous thermo and light cycles, reaching the peak of TIM staining at the beginning of the dark phase, also corresponding to the middle of the cryophase (Miyasako *et al.*, 2007). These observations suggest that DN2s and LPNs might be particularly important for temperature entrainment. In support of this hypothesis, Busza and coworkers have also found that although the s-LNvs were able to entrain to thermocycles even in the absence of other functional clock cells, they were not required for, and actually slowed down, the temperature entrainment of activity rhythms (Busza *et al.*, 2007). On the contrary, CRY-negative cells, such as DN2s and LPNs, appeared to play a prominent role in these conditions. Furthermore, Picot and colleagues showed that in third-instar larvae, the DN2s were the only neurons with identical phase of PER and TIM oscillation under light or temperature entrainment, and that their light entrainment but not their temperature entrainment required signaling from the s-LNvs (Picot *et al.*, 2009).

These results advocate cell-specific responses to temperature but are not informative on the pathways required. An implicit assumption of temperature entrainment has been that temperature affects the molecular clock directly and identically in all clock cells, as opposed to light, which requires dedicated input pathways. However, the observations of Sehadova and coworkers have proved this belief wrong (Sehadova *et al.*, 2009). Using a luciferase reporter system, they showed that whole heads and other body parts (legs, wings, etc.,) could entrain to temperature cycles (as judged from whole-tissue luciferase oscillations), although isolated brains could not. This suggests that the brain depends upon signals from the periphery for temperature entrainment to occur. Previous work (Glaser and Stanewsky, 2005) had identified *nocte* as a gene required for temperature entrainment of locomotor activity rhythms. The cloning of the gene revealed that *nocte* encodes for a putative transcription regulator expressed in the peripheral nervous system and in the chordotonal organs in particular (although some expression was also detected in the brain). The protein product of *nocte* was proved essential for the proper morphophysiological function of these structures, suggesting that they are involved in temperature entrainment. In fact, other mutants also affecting the morphology and the physiology of chordotonal organs equally showed a deficit in temperature entrainment. The chordotonal organs are stretch receptors internally attached to the cuticle and located at the joints between segments. They mediate proprioception (legs and wings), gravireception (antennae), and vibration detection (antennae) (Kernan, 2007). Since luciferase oscillations showed that whole heads could entrain to thermocycles in culture, but surgical removal of the antennae did not affect temperature entrainment in flies, the chordotonal organ in legs and wings must be sufficient for temperature entrainment, and therefore, project either directly or indirectly to the clock neurons in the brain (Sehadova *et al.*, 2009).

Other observations further suggest that temperature entrainment is more complicated than previously thought. For instance, the robust behavioral rhythms observed with temperature cycles under LL, suggest an overlap between the molecular mechanisms controlling light and temperature entrainment. Interestingly, the phospholipase C encoded by the *no-receptor-potential-A* (*norpA*) gene, which is required for the phototransduction pathway in photoreceptor cells, is involved in both light (Stanewsky *et al.*, 1998) and temperature (Glaser and Stanewsky, 2005) entrainment. Loss of function mutants *norpA*^{P24} and *norpA*^{P41} did not entrain to thermocycles *in vitro* (whole-tissue luciferase oscillations) or *in vivo* (entrainment of rhythmic locomotor activity in flies) (Glaser and Stanewsky, 2005). However, there are not known mechanisms able to explain the molecular requirement for NORPA in thermosensing, including NORPA involvement in the thermal regulation of a splicing event in the 3′ UTR of *per*. In wild-type flies, both cold temperatures and short photoperiod increase the level of this splicing event, resulting in earlier accumulation of *per* mRNA (Collins *et al.*, 2004; Majercak *et al.*, 2004). This correlates with a steeper increase in PER levels and in an earlier locomotor activity bout, now peaking in the middle rather than at the end of the day. In *norpA* mutants, *per* spliced forms are constitutively high regardless of ambient temperature. However, thermocycles produce equal amounts of spliced and nonspliced forms of *per* in wild-type flies, both under DD or LL, ruling out the involvement of this mechanism in temperature entrainment (Glaser and Stanewsky, 2005).

IV. WHICH CLOCK IS IT ANYWAY?

Circadian rhythmicity is found in all natural kingdoms and, although the molecular components are not conserved across the main divisions of life, all clocks are seemingly based on the same design where TTLs appear to impart 24 h cycles to cellular functions (Dunlap, 1999). The latter assumption has been challenged in recent years. The first surprise was the discovery that rhythmic autophosphorylation of the cyanobacteria clock protein KaiC does not demand cyclic RNA and protein expression (Tomita *et al.*, 2005), actually, cyclic KaiC autophosphorylation persists in the test tube requiring only the presence of ATP and of the two other clock proteins, KaiA and KaiB (Nakajima *et al.*, 2005). Then, several experiments in flies and mammals showed that constitutive over-expression of circadian proteins does not impede behavioral (Numano *et al.*, 2006; Yang and Sehgal, 2001) or cellular rhythmicity (Yamanaka *et al.*, 2007). Equally, protein translation is not required for circadian rhythmicity as SCN slices or retinas of the sea snail *Bulla gouldiana* generally maintain their rhythmic phase and periodicity after exposure to the translation inhibitor cycloheximide (Khalsa *et al.*, 1992; Yamaguchi *et al.*, 2003). The implication is that preexisting proteins can sustain circadian rhythmicity forming an enzymatic clock that is coupled to but is not strictly dependent upon the TTLs. To confirm this prediction, two studies have used a posttranslational biomarker for rhythmicity, namely the cycle of oxidation/reduction of peroxiredoxin (PRX) proteins. PRXs are a highly conserved family of antioxidant proteins involved in the control of intracellular peroxide levels. A circadian oscillation of PXRs oxidation was demonstrated in human erythrocytes, anucleated cells that have lost any transcriptional capability (O'Neill and Reddy, 2011), and in the unicellular picoeukaryotic alga *Ostreococcus tauri* in the presence of inhibitors of transcription (cordycepin) and translation (cycloheximide) (O'Neill et al., 2011). Moreover, circadian oscillations of other cytoplasmic parameters, such as hemoglobin tetramer–dimer transitions and NADH/NADPH oscillations, were shown in erythrocytes (O'Neill and Reddy, 2011). This is consistent with the view that circadian modulation via an enzymatic clock does not require the nucleus (Lakin-Thomas, 2006), as first shown in *Acetabularia crenulata* (a green alga) cells, which maintain circadian photosynthetic rhythms even when anucleated (Sweeney and Haxo, 1961). The presence of an enzymatic clock coupled to the TTLs is also suggested by the fact that small cytoplasmic signaling molecules, such as Ca^{2+} and cAMP that are typically considered output of the TTLs, cycle in anticipation and control the expression of canonical clock genes, which on the one hand suggests they are under control of a parallel cycling system, but on the other it elevates them to the role of fundamental components of the TTLs *per se* (for a review, see Hastings et al., 2008). We can then envisage a dual-components model where an enzymatic clock is interconnected with the TTLs in virtue of sharing common elements (small signaling molecules, kinases, phosphatases, etc.,), which are the final effectors of cellular rhythmicity (Hastings et al., 2008). If we accept that it is the cyclic activity (which not necessarily converts into cyclic amount) of these effectors what really matters for circadian rhythmicity, we can then rationalize some observations that are difficult to explain under the TTLs model alone. For instance, assuming that the TTLs are the sole driver of rhythmicity, cyclic posttranslational modifications should not take place, and flies should become arrhythmic when PER or TIM or both are held constant, but this is not invariably the case. Rather, the outcome depends upon the overall level of protein expression, with arrhythmicity developing as a consequence of high PER, TIM, or PER and TIM production (Yang and Sehgal, 2001). This suggests that another cycling mechanism must be operating to regulate the function of the kinases/phosphatases/E3 ligases that control PER and TIM posttranslational modifications, and that the rhythmic activity of those enzymes is suppressed by an excess of substrate. Therefore, the main function of the TTLs might actually lay in providing cycling substrates (such as PER, TIM, and CLK), able to divert the function of essential enzymes (as those involved in the enzymatic clock) from crucial metabolic pathways to futile (in metabolic terms) cycles with the sole but important function of increasing the robustness and amplitude of the enzymatic clock. Under this scenario, the kinases/phosphatases/E3 ligases mentioned above become the core of the cycling mechanisms that constitute the clock rather than being accessories to the TTLs. In particular, the current TTLs model suggests that key cycling components, such as PER, TIM, and CLK, generate rhythmicity by driving rhythmic transcription of ccgs, and that the main role of kinases/phosphatases/E3 ligases is to increase and delay the cycling of those key players. However, we now know that rhythmic mRNAs (and there are many of them) do not necessarily give rise to rhythmic proteins, and equally the latter do not require the former (Akhtar *et al.*, 2002; Reddy *et al.*, 2006), making unsustainable the claim that rhythmicity is ultimately driven by rhythmic regulators of transcription. Instead, rhythmic enzymatic activity is what finally matters to generate rhythmic phenotypes. We can then see how rhythmic elements of the TTLs, such as PER, TIM, and CLK, might contribute to cellular cycles by segregating a considerable portion of the cellular pool of important kinases/phosphatases/E3 ligase, etc., at certain times during the day, thus controlling the general activity of the cell rather than transcription alone (Fig. 4.10).

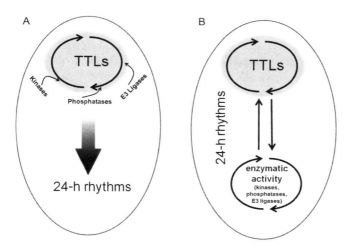

Figure 4.10. Two models of circadian regulation. (A) The established model of the clock suggests that rhythmic transcription/translation feedback loops (TTLs) are ultimately responsible for driving circadian cycles in the cell. Enzymes such as kinases, phosphatases, E3-ubiquitin ligases, etc., are accessories to the TTLs. Their function is to stabilize the cycling of important circadian proteins (such as PER, TIM, and CLK) which are ultimately responsible for generating 24 h rhythmicity. (B) A dual-component model where kinases, phosphatases, E3-ubiquitin ligases, etc., constitute an enzymatic clock that is interconnected with the TTLs. Cycling enzymatic activity is ultimately responsible for 24 h rhythmicity (as evidenced by blocking rhythmic transcription/translation) but robust and long-lasting rhythms require functioning TTLs. (See Color Insert.)

V. CONCLUSIONS

We would like to finish by offering our personal vision of the circadian system and of its evolution. Our hypothesis (see also Lakin-Thomas, 2006) is that "rhythmicity" is a general property of the cell, embedded in the metabolic processes. However, to channel that diffused rhythmicity into discrete pheno-types, natural selection must have favored the evolution of an additional system, synergistic to the first, resulting in high-amplitude rhythms of enzymatic activity. Rhythmic transcription, probably not a determinant of rhythmicity in the first place, would emerge as the inevitable consequence of the rhythmic activity of the enzymatic clock, which is composed of multifunctional regulatory proteins (kinases, phosphatase, E3 ligases, proteasome, and associated components) that also control transcription factors. Hence, the "rhythmic transcriptome" is prob-ably more a reflection of the cell-specific complement of transcription factors rather than being causal to the nature of the clock in different tissues. That would explain why so many genes are rhythmically transcribed, and why there is so much variability in the type of cycling genes among tissues (Akhtar et al., 2002; Miller et al., 2007; Storch et al., 2002). Having rhythmic enzymatic activity on the one hand and emerging rhythmic transcription on the other, the need to match the phases of these oscillations would then arise. These adjustments would also require posttranscriptional and posttranslational modifications, which meant tinkering with the enzymatic clock to harmonize its functioning with the emerg-ing rhythms in gene expression. The following step to further stabilize these interactions might have been to recruit genes and proteins into a new function, the control of rhythmicity per se, based upon their ability to temporally segregate the activity of the components of the enzymatic clock. Thus, we speculate that circadian rhythmicity, which was previously a diffuse cellular property, evolved into a specialized cellular function. The dominance of the LD cycle in regulating daily functions would explain why both molecularly and anatomically clocks have a strong association with photopigments and photoreceptors.

We further hypothesize that in animals, an additional level of regula-tion would emerge in the brain, the organ exerting centralized control over the rest of the body. The fundamental properties of the clock such as self-sustained rhythmicity, entrainability, and temperature compensation already appear in single cells, but these are quite limited; coupling several oscillators greatly strengthens the clock, and this can be demonstrated even in artificial systems (Danino et al., 2010). Moreover, coupling oscillators that have different phases (for instance, in the mouse SCN the dorsomedial area phase-advances the ventrolateral region; Yamaguchi et al., 2003) or even periods (in Drosophila, the DN2s have a period about 2 h shorter than the other neurons; Nitabach et al., 2006) would produce greater plasticity, as changes in phase or period of the overall clock would simply require a shift by the different oscillators in the weight of their contributions to the network. The importance of this is, for instance, facilitating adaptation to the changes in photoperiod, and temperature experienced in different seasons or adjusting the clock in response to social stimuli (Levine *et al.*, 2002; Scheibler and Wollnik, 2009; Shafer *et al.*, 2004). We would then expect that in neurons, the evolution of the clock would favor many modes of intercellular communication, and that coupling would acquire paramount importance to the point of resulting in less robust cell-autonomous oscillators. That would generate a system able to operate as a highly plastic network with distinctive properties that go beyond those of individual neurons. These predictions are met by the mammalian system, where the SCN can maintain precise and indefinite circadian rhythmicity when maintained in organotypic slice cultures whereas isolated cells reveal unstable and noisy circadian oscillations (Webb *et al.*, 2009). On the contrary, the humble fibroblast is well equipped for maintaining robust circadian oscillations in isolation (Welsh *et al.*, 2004). Although analogous *in vitro* experiments have not been performed in *Drosophila*, the dependence on PDF signaling *in vivo* to maintain cycling amplitude and synchronization (Lin *et al.*, 2004), suggests that the brain clock in the fly might be equally dependent on network interactions (Fig. 4.11).

It is an exciting time for chronobiology as new challenges fill the horizon. The most relevant will be to understand the logic that rules the enzymatic clock and to unravel the emerging properties of a network of oscillating neurons. We envisage that the main hurdle on this path will be the resistance

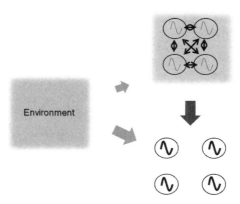

Figure 4.11. The clock in the brain is specialized for coupling. In the periphery, cells exhibit a robust clock that is largely capable of independent cycling. These cells are strongly dependent from inputs, from the brain and/or the environment to synchronize. Central clock cells have traded cell-autonomous self-sustained rhythmicity in favor of coupling. This gives rise to an oscillating network with properties that go well beyond those of single cells in terms of endurance, plasticity, and responsiveness. (See Color Insert.)

in letting go of a simple and elegant model that has been useful and productive for over 30 years. The amount of knowledge produced under the rule of the TTLs is vast, but we are reaching the boundaries of its worth. We look forward to the establishment of a more general model in the years to come.

Acknowledgments

We acknowledge funding and support from the BBSRC, NERC, MRC, and NC3Rs. We declare that there are not conflicts of interest. We apologize to colleagues whose work has not been cited due to space constrain.

References

Ahmad, M., and Cashmore, A. R. (1996). Seeing blue: The discovery of cryptochrome. *Plant Mol. Biol.* **30,** 851–861.

Akhtar, R. A., Reddy, A. B., Maywood, E. S., Clayton, J. D., King, V. M., Smith, A. G., Gant, T. W., Hastings, M. H., and Kyriacou, C. P. (2002). Circadian cycling of the mouse liver transcriptome, as revealed by cDNA microarray, is driven by the suprachiasmatic nucleus. *Curr. Biol.* **12,** 540–550.

Akten, B., Jauch, E., Genova, G. K., Kim, E. Y., Edery, I., Raabe, T., and Jackson, F. R. (2003). A role for CK2 in the Drosophila circadian oscillator. *Nat. Neurosci.* **6,** 251–257.

Allada, R., White, N. E., So, W. V., Hall, J. C., and Rosbash, M. (1998). A mutant drosophila homolog of mammalian clock disrupts circadian rhythms and transcription of period and timeless. *Cell* **93,** 791–804.

Ashmore, L. J., Sathyanarayanan, S., Silvestre, D. W., Emerson, M. M., Schotland, P., and Sehgal, A. (2003). Novel insights into the regulation of the timeless protein. *J. Neurosci.* **23,** 7810–7819.

Bae, K., Lee, C., Hardin, P. E., and Edery, I. (2000). DCLOCK is present in limiting amounts and likely mediates daily interactions between the dCLOCK-CYC transcription factor and the PER-TIM complex. *J. Neurosci.* **20,** 1746–1753.

Bao, S., Rihel, J., Bjes, E., Fan, J. Y., and Price, J. L. (2001). The Drosophila double-timeS mutation delays the nuclear accumulation of period protein and affects the feedback regulation of period mRNA. *J. Neurosci.* **21,** 7117–7126.

Baylies, M. K., Bargiello, T. A., Jackson, F. R., and Young, M. W. (1987). Changes in abundance or structure of the per gene product can alter periodicity of the Drosophila clock. *Nature* **326,** 390–392.

Baylies, M. K., Vosshall, L. B., Sehgal, A., and Young, M. W. (1992). New short period mutations of the Drosophila clock gene per. *Neuron* **9,** 575–581.

Belvin, M. P., Zhou, H., and Yin, J. C. (1999). The Drosophila dCREB2 gene affects the circadian clock. *Neuron* **22,** 777–787.

Benito, J., Zheng, H., and Hardin, P. E. (2007). PDP1epsilon functions downstream of the circadian oscillator to mediate behavioral rhythms. *J. Neurosci.* **27,** 2539–2547.

Benito, J., Houl, J. H., Roman, G. W., and Hardin, P. E. (2008). The blue-light photoreceptor CRYPTOCHROME is expressed in a subset of circadian oscillator neurons in the Drosophila CNS. *J. Biol. Rhythms* **23,** 296–307.

Blanchardon, E., Grima, B., Klarsfeld, A., Chelot, E., Hardin, P. E., Preat, T., and Rouyer, F. (2001). Defining the role of Drosophila lateral neurons in the control of circadian rhythms in motor activity and Eclosion by targeted genetic ablation and PERIOD protein overexpression. *Eur. J. Neurosci.* **13**, 871–888.

Blau, J., and Young, M. W. (1999). Cycling vrille expression is required for a functional Drosophila clock. *Cell* **99**, 661–671.

Busza, A., Emery-Le, M., Rosbash, M., and Emery, P. (2004). Roles of the two drosophila CRYPTO-CHROME structural domains in circadian photoreception. *Science* **304**, 1503–1506.

Busza, A., Murad, A., and Emery, P. (2007). Interactions between circadian neurons control temperature synchronization of Drosophila behavior. *J. Neurosci.* **27**, 10722–10733.

Ceriani, M. F., Darlington, T. K., Staknis, D., Mas, P., Petti, A. A., Weitz, C. J., and Kay, S. A. (1999). Light-dependent sequestration of TIMELESS by CRYPTOCHROME. *Science* **285**, 553–556.

Ceriani, M. F., Hogenesch, J. B., Yanovsky, M., Panda, S., Straume, M., and Kay, S. A. (2002). Genome-wide expression analysis in Drosophila reveals genes controlling circadian behavior. *J. Neurosci.* **22**, 9305–9319.

Chang, D. C., and Reppert, S. M. (2003). A novel C-terminal domain of Drosophila PERIOD inhibits dCLOCK:CYCLE-mediated transcription. *Curr. Biol.* **13**, 758–762.

Chen, K. F., Peschel, N., Zavodska, R., Sehadova, H., and Stanewsky, R. (2011). QUASIMODO, a novel GPI-anchored zona pellucida protein involved in light input to the Drosophila circadian clock. *Curr. Biol.* **21**, 719–729.

Chiu, J. C., Vanselow, J. T., Kramer, A., and Edery, I. (2008). The phospho-occupancy of an atypical SLIMB-binding site on PERIOD that is phosphorylated by DOUBLETIME controls the pace of the clock. *Genes Dev.* **22**, 1758–1772.

Chiu, J. C., Ko, H. W., and Edery, I. (2011). NEMO/NLK phosphorylates PERIOD to initiate a time-delay phosphorylation circuit that sets circadian clock speed. *Cell* **145**, 357–370.

Claridge-Chang, A., Wijnen, H., Naef, F., Boothroyd, C., Rajewsky, N., and Young, M. W. (2001). Circadian regulation of gene expression systems in the Drosophila head. *Neuron* **32**, 657–671.

Collins, B. H., Rosato, E., and Kyriacou, C. P. (2004). Seasonal behavior in Drosophila melanogaster requires the photoreceptors, the circadian clock, and phospholipase C. *Proc. Natl. Acad. Sci. U.S.A.* **101**, 1945–1950.

Cyran, S. A., Buchsbaum, A. M., Reddy, K. L., Lin, M. C., Glossop, N. R., Hardin, P. E., Young, M. W., Storti, R. V., and Blau, J. (2003). Vrille, Pdp1, and dClock form a second feedback loop in the Drosophila circadian clock. *Cell* **112**, 329–341.

Cyran, S. A., Yiannoulos, G., Buchsbaum, A. M., Saez, L., Young, M. W., and Blau, J. (2005). The double-time protein kinase regulates the subcellular localization of the Drosophila clock protein period. *J. Neurosci.* **25**, 5430–5437.

Danino, T., Mondragon-Palomino, O., Tsimring, L., and Hasty, J. (2010). A synchronized quorum of genetic clocks. *Nature* **463**, 326–330.

Dunlap, J. C. (1999). Molecular bases for circadian clocks. *Cell* **96**, 271–290.

Dushay, M. S., Rosbash, M., and Hall, J. C. (1989). The disconnected visual system mutations in Drosophila Melanogaster drastically disrupt circadian rhythms. *J. Biol. Rhythms* **4**, 1–27.

Edery, I., Rutila, J. E., and Rosbash, M. (1994). Phase shifting of the circadian clock by induction of the Drosophila period protein. *Science* **263**, 237–240.

Emery, P., So, W. V., Kaneko, M., Hall, J. C., and Rosbash, M. (1998). CRY, a Drosophila clock and light-regulated cryptochrome, is a major contributor to circadian rhythm resetting and photosensitivity. *Cell* **95**, 669–679.

Emery, P., Stanewsky, R., Hall, J. C., and Rosbash, M. (2000a). A unique circadian-rhythm photoreceptor. *Nature* **404**, 456–457.

Emery, P., Stanewsky, R., Helfrich-Forster, C., Emery-Le, M., Hall, J. C., and Rosbash, M. (2000b). Drosophila CRY is a deep brain circadian photoreceptor. *Neuron* **26**, 493–504.

Ewer, J., Frisch, B., Hamblen-Coyle, M. J., Rosbash, M., and Hall, J. C. (1992). Expression of the period clock gene within different cell types in the brain of Drosophila adults and mosaic analysis of these cells' influence on circadian behavioral rhythms. *J. Neurosci.* **12**, 3321–3349.

Fang, Y., Sathyanarayanan, S., and Sehgal, A. (2007). Post-translational regulation of the Drosophila circadian clock requires protein phosphatase 1 (PP1). *Genes Dev.* **21**, 1506–1518.

Fogle, K. J., Parson, K. G., Dahm, N. A., and Holmes, T. C. (2011). CRYPTOCHROME is a blue-light sensor that regulates neuronal firing rate. *Science* **331**, 1409–1413.

Frisch, B., Hardin, P. E., Hamblen-Coyle, M. J., Rosbash, M., and Hall, J. C. (1994). A promoterless period gene mediates behavioral rhythmicity and cyclical per expression in a restricted subset of the Drosophila nervous system. *Neuron* **12**, 555–570.

Fujii, S., and Amrein, H. (2010). Ventral lateral and DN1 clock neurons mediate distinct properties of male sex drive rhythm in Drosophila. *Proc. Natl. Acad. Sci. U.S.A.* **107**, 10590–10595.

Gekakis, N., Saez, L., Delahaye-Brown, A. M., Myers, M. P., Sehgal, A., Young, M. W., and Weitz, C. J. (1995). Isolation of timeless by PER protein interaction: Defective interaction between timeless protein and long-period mutant PERL. *Science* **270**, 811–815.

Glaser, F. T., and Stanewsky, R. (2005). Temperature synchronization of the Drosophila circadian clock. *Curr. Biol.* **15**, 1352–1363.

Glossop, N. R., Lyons, L. C., and Hardin, P. E. (1999). Interlocked feedback loops within the Drosophila circadian oscillator. *Science* **286**, 766–768.

Glossop, N. R., Houl, J. H., Zheng, H., Ng, F. S., Dudek, S. M., and Hardin, P. E. (2003). VRILLE feeds back to control circadian transcription of clock in the Drosophila circadian oscillator. *Neuron* **37**, 249–261.

Grima, B., Lamouroux, A., Chelot, E., Papin, C., Limbourg-Bouchon, B., and Rouyer, F. (2002). The F-box protein slimb controls the levels of clock proteins period and timeless. *Nature* **420**, 178–182.

Grima, B., Chelot, E., Xia, R., and Rouyer, F. (2004). Morning and evening peaks of activity rely on different clock neurons of the Drosophila brain. *Nature* **431**, 869–873.

Gummadova, J. O., Coutts, G. A., and Glossop, N. R. (2009). Analysis of the Drosophila clock promoter reveals heterogeneity in expression between subgroups of central oscillator cells and identifies a novel enhancer region. *J. Biol. Rhythms* **24**, 353–367.

Handler, A. M., and Konopka, R. J. (1979). Transplantation of a circadian pacemaker in Drosophila. *Nature* **279**, 236–238.

Hardin, P. E., Hall, J. C., and Rosbash, M. (1992). Circadian oscillations in period gene mRNA levels are transcriptionally regulated. *Proc. Natl. Acad. Sci. U.S.A.* **89**, 11711–11715.

Harmar, A. J., Marston, H. M., Shen, S., Spratt, C., West, K. M., Sheward, W. J., Morrison, C. F., Dorin, J. R., Piggins, H. D., Reubi, J. C., et al. (2002). The VPAC(2) receptor is essential for circadian function in the mouse suprachiasmatic nuclei. *Cell* **109**, 497–508.

Hastings, M. H., Maywood, E. S., and O'Neill, J. S. (2008). Cellular circadian pacemaking and the role of cytosolic rhythms. *Curr. Biol.* **18**, R805–R815.

Helfrich, C. (1986). Role of the optic lobes in the regulation of the locomotor activity rhythm of Drosophila melanogaster: Behavioral analysis of neural mutants. *J. Neurogenet.* **3**, 321–343.

Helfrich, C., and Engelmann, W. (1983). Circadian rhythms of the locomotor activity in *Drosophila melanogaster* and its mutants 'Sine Oculis' and 'Small Optic Lobes'. *Physiol. Entomol.* **8**, 257–272.

Helfrich-Forster, C. (1998). Robust circadian rhythmicity of Drosophila melanogaster requires the presence of lateral neurons: A brain-behavioral study of disconnected mutants. *J. Comp. Physiol. A* **182**, 435–453.

Helfrich-Forster, C. (2003). The neuroarchitecture of the circadian clock in the brain of Drosophila melanogaster. *Microsc. Res. Tech.* **62**, 94–102.

Helfrich-Forster, C. (2005). Techniques that revealed the network of the circadian clock of Drosophila. *Methods Enzymol.* **393**, 439–451.

Helfrich-Forster, C., Winter, C., Hofbauer, A., Hall, J. C., and Stanewsky, R. (2001). The circadian clock of fruit flies is blind after elimination of all known photoreceptors. *Neuron* **30**, 249–261.

Hermann, C., Yoshii, T., Dusik, V., and Helfrich-Forster, C. (2012). The neuropeptide F immunoreactive clock neurons modify evening locomotor activity and free-running period in Drosophila melanogaster. *J. Comp. Neurol.* **520**, 970–987.

Huang, Z. J., Edery, I., and Rosbash, M. (1993). PAS is a dimerization domain common to Drosophila period and several transcription factors. *Nature* **364**, 259–262.

Hung, H. C., Maurer, C., Kay, S. A., and Weber, F. (2007). Circadian transcription depends on limiting amounts of the transcription co-activator nejire/CBP. *J. Biol. Chem.* **282**, 31349–31357.

Hunter-Ensor, M., Ousley, A., and Sehgal, A. (1996). Regulation of the Drosophila protein timeless suggests a mechanism for resetting the circadian clock by light. *Cell* **84**, 677–685.

Kadener, S., Stoleru, D., McDonald, M., Nawathean, P., and Rosbash, M. (2007). Clockwork orange is a transcriptional repressor and a new Drosophila circadian pacemaker component. *Genes Dev.* **21**, 1675–1686.

Kaneko, M., and Hall, J. C. (2000). Neuroanatomy of cells expressing clock genes in Drosophila: Transgenic manipulation of the period and timeless genes to mark the Perikarya of circadian pacemaker neurons and their projections. *J. Comp. Neurol.* **422**, 66–94.

Kaneko, M., Helfrich-Forster, C., and Hall, J. C. (1997). Spatial and temporal expression of the period and timeless genes in the developing nervous system of Drosophila: Newly identified pacemaker candidates and novel features of clock gene product cycling. *J. Neurosci.* **17**, 6745–6760.

Kernan, M. J. (2007). Mechanotransduction and auditory transduction in Drosophila. *Pflugers Arch.* **454**, 703–720.

Khalsa, S. B., Whitmore, D., and Block, G. D. (1992). Stopping the circadian pacemaker with inhibitors of protein synthesis. *Proc. Natl. Acad. Sci. U.S.A.* **89**, 10862–10866.

Kim, E. Y., Ko, H. W., Yu, W., Hardin, P. E., and Edery, I. (2007). A DOUBLETIME kinase binding domain on the Drosophila PERIOD protein is essential for its hyperphosphorylation, transcriptional repression, and circadian clock function. *Mol. Cell. Biol.* **27**, 5014–5028.

Kloss, B., Price, J. L., Saez, L., Blau, J., Rothenfluh, A., Wesley, C. S., and Young, M. W. (1998). The Drosophila clock gene double-time encodes a protein closely related to human casein kinase Iepsilon. *Cell* **94**, 97–107.

Kloss, B., Rothenfluh, A., Young, M. W., and Saez, L. (2001). Phosphorylation of period is influenced by cycling physical associations of double-time, period, and timeless in the Drosophila clock. *Neuron* **30**, 699–706.

Ko, H. W., Jiang, J., and Edery, I. (2002). Role for slimb in the degradation of Drosophila period protein phosphorylated by doubletime. *Nature* **420**, 673–678.

Ko, H. W., Kim, E. Y., Chiu, J., Vanselow, J. T., Kramer, A., and Edery, I. (2010). A hierarchical phosphorylation cascade that regulates the timing of PERIOD nuclear entry reveals novel roles for proline-directed kinases and GSK-3beta/SGG in circadian clocks. *J. Neurosci.* **30**, 12664–12675.

Koh, K., Zheng, X., and Sehgal, A. (2006). JETLAG resets the Drosophila circadian clock by promoting light-induced degradation of TIMELESS. *Science* **312**, 1809–1812.

Konopka, R. J., and Benzer, S. (1971). Clock mutants of Drosophila melanogaster. *Proc. Natl. Acad. Sci. U.S.A.* **68**, 2112–2116.

Konopka, R., Wells, S., and Lee, T. (1983). Mosaic analysis of a *Drosophila* clock mutant. *Mol. Gen. Genet.* **190**, 284–288.

Kula-Eversole, E., Nagoshi, E., Shang, Y., Rodriguez, J., Allada, R., and Rosbash, M. (2010). Surprising gene expression patterns within and between PDF-containing circadian neurons in Drosophila. *Proc. Natl. Acad. Sci. U.S.A.* **107**, 13497–13502.

Kwok, R. P., Lundblad, J. R., Chrivia, J. C., Richards, J. P., Bachinger, H. P., Brennan, R. G., Roberts, S. G., Green, M. R., and Goodman, R. H. (1994). Nuclear protein CBP is a coactivator for the transcription factor CREB. *Nature* **370,** 223–226.

Lakin-Thomas, P. L. (2006). Transcriptional feedback oscillators: Maybe, maybe not. *J. Biol. Rhythms* **21,** 83–92.

Lamaze, A., Lamouroux, A., Vias, C., Hung, H. C., Weber, F., and Rouyer, F. (2011). The E3 ubiquitin ligase CTRIP controls CLOCK levels and PERIOD oscillations in Drosophila. *EMBO Rep.* **12,** 549–557.

Lee, C., Bae, K., and Edery, I. (1998). The Drosophila CLOCK protein undergoes daily rhythms in abundance, phosphorylation, and interactions with the PER-TIM complex. *Neuron* **21,** 857–867.

Lee, C., Bae, K., and Edery, I. (1999). PER and TIM inhibit the DNA binding activity of a Drosophila CLOCK-CYC/dBMAL1 heterodimer without disrupting formation of the heterodimer: A basis for circadian transcription. *Mol. Cell. Biol.* **19,** 5316–5325.

Levine, J. D., Funes, P., Dowse, H. B., and Hall, J. C. (2002). Resetting the circadian clock by social experience in Drosophila melanogaster. *Science* **298,** 2010–2012.

Lim, C., Lee, J., Choi, C., Kim, J., Doh, E., and Choe, J. (2007a). Functional role of CREB-binding protein in the circadian clock system of Drosophila melanogaster. *Mol. Cell. Biol.* **27,** 4876–4890.

Lim, C., Lee, J., Koo, E., and Choe, J. (2007b). Targeted inhibition of Pdp1epsilon abolishes the circadian behavior of Drosophila melanogaster. *Biochem. Biophys. Res. Commun.* **364,** 294–300.

Lim, C., Chung, B. Y., Pitman, J. L., McGill, J. J., Pradhan, S., Lee, J., Keegan, K. P., Choe, J., and Allada, R. (2007c). Clockwork orange encodes a transcriptional repressor important for circadian-clock amplitude in Drosophila. *Curr. Biol.* **17,** 1082–1089.

Lim, C., Lee, J., Choi, C., Kilman, V. L., Kim, J., Park, S. M., Jang, S. K., Allada, R., and Choe, J. (2011). The novel gene twenty-four defines a critical translational step in the Drosophila clock. *Nature* **470,** 399–403.

Lin, J. M., Kilman, V. L., Keegan, K., Paddock, B., Emery-Le, M., Rosbash, M., and Allada, R. (2002). A role for casein kinase 2alpha in the Drosophila circadian clock. *Nature* **420,** 816–820.

Lin, Y., Stormo, G. D., and Taghert, P. H. (2004). The neuropeptide pigment-dispersing factor coordinates pacemaker interactions in the Drosophila circadian system. *J. Neurosci.* **24,** 7951–7957.

Lin, J. M., Schroeder, A., and Allada, R. (2005). In vivo circadian function of casein kinase 2 phosphorylation sites in Drosophila PERIOD. *J. Neurosci.* **25,** 11175–11183.

Lopez-Molina, L., Conquet, F., Dubois-Dauphin, M., and Schibler, U. (1997). The DBP gene is expressed according to a circadian rhythm in the suprachiasmatic nucleus and influences circadian behavior. *EMBO J.* **16,** 6762–6771.

Majercak, J., Chen, W. F., and Edery, I. (2004). Splicing of the period gene 3'-terminal intron is regulated by light, circadian clock factors, and phospholipase C. *Mol. Cell. Biol.* **24,** 3359–3372.

Marrus, S. B., Zeng, H., and Rosbash, M. (1996). Effect of constant light and circadian entrainment of perS flies: Evidence for light-mediated delay of the negative feedback loop in Drosophila. *EMBO J.* **15,** 6877–6886.

Martinek, S., Inonog, S., Manoukian, A. S., and Young, M. W. (2001). A role for the segment polarity gene shaggy/GSK-3 in the Drosophila circadian clock. *Cell* **105,** 769–779.

Matsumoto, A., Ukai-Tadenuma, M., Yamada, R. G., Houl, J., Uno, K. D., Kasukawa, T., Dauwalder, B., Itoh, T. Q., Takahashi, K., Ueda, R., *et al.* (2007). A functional genomics strategy reveals clockwork orange as a transcriptional regulator in the Drosophila circadian clock. *Genes Dev.* **21,** 1687–1700.

Maywood, E. S., Reddy, A. B., Wong, G. K., O'Neill, J. S., O'Brien, J. A., McMahon, D. G., Harmar, A. J., Okamura, H., and Hastings, M. H. (2006). Synchronization and maintenance of timekeeping in suprachiasmatic circadian clock cells by neuropeptidergic signaling. *Curr. Biol.* **16,** 599–605.

McDonald, M. J., and Rosbash, M. (2001). Microarray analysis and organization of circadian gene expression in Drosophila. *Cell* **107**, 567–578.

Meissner, R. A., Kilman, V. L., Lin, J. M., and Allada, R. (2008). TIMELESS is an important mediator of CK2 effects on circadian clock function in vivo. *J. Neurosci.* **28**, 9732–9740.

Menet, J. S., Abruzzi, K. C., Desrochers, J., Rodriguez, J., and Rosbash, M. (2010). Dynamic PER repression mechanisms in the Drosophila circadian clock: From on-DNA to off-DNA. *Genes Dev.* **24**, 358–367.

Meyer, P., Saez, L., and Young, M. W. (2006). PER-TIM interactions in living Drosophila cells: An interval timer for the circadian clock. *Science* **311**, 226–229.

Miller, B. H., McDearmon, E. L., Panda, S., Hayes, K. R., Zhang, J., Andrews, J. L., Antoch, M. P., Walker, J. R., Esser, K. A., Hogenesch, J. B., et al. (2007). Circadian and CLOCK-controlled regulation of the mouse transcriptome and cell proliferation. *Proc. Natl. Acad. Sci. U.S.A.* **104**, 3342–3347.

Mitsui, S., Yamaguchi, S., Matsuo, T., Ishida, Y., and Okamura, H. (2001). Antagonistic role of E4BP4 and PAR proteins in the circadian oscillatory mechanism. *Genes Dev.* **15**, 995–1006.

Miyasako, Y., Umezaki, Y., and Tomioka, K. (2007). Separate sets of cerebral clock neurons are responsible for light and temperature entrainment of Drosophila circadian locomotor rhythms. *J. Biol. Rhythms* **22**, 115–126.

Myers, M. P., Wager-Smith, K., Wesley, C. S., Young, M. W., and Sehgal, A. (1995). Positional cloning and sequence analysis of the Drosophila clock gene, timeless. *Science* **270**, 805–808.

Myers, M. P., Wager-Smith, K., Rothenfluh-Hilfiker, A., and Young, M. W. (1996). Light-induced degradation of TIMELESS and entrainment of the Drosophila circadian clock. *Science* **271**, 1736–1740.

Nagoshi, E., Sugino, K., Kula, E., Okazaki, E., Tachibana, T., Nelson, S., and Rosbash, M. (2010). Dissecting differential gene expression within the circadian neuronal circuit of Drosophila. *Nat. Neurosci.* **13**, 60–68.

Naidoo, N., Song, W., Hunter-Ensor, M., and Sehgal, A. (1999). A role for the proteasome in the light response of the timeless clock protein. *Science* **285**, 1737–1741.

Nakajima, M., Imai, K., Ito, H., Nishiwaki, T., Murayama, Y., Iwasaki, H., Oyama, T., and Kondo, T. (2005). Reconstitution of circadian oscillation of cyanobacterial KaiC phosphorylation in vitro. *Science* **308**, 414–415.

Ng, F. S., Tangredi, M. M., and Jackson, F. R. (2011). Glial cells physiologically modulate clock neurons and circadian behavior in a calcium-dependent manner. *Curr. Biol.* **21**, 625–634.

Nitabach, M. N., Blau, J., and Holmes, T. C. (2002). Electrical silencing of Drosophila pacemaker neurons stops the free-running circadian clock. *Cell* **109**, 485–495.

Nitabach, M. N., Wu, Y., Sheeba, V., Lemon, W. C., Strumbos, J., Zelensky, P. K., White, B. H., and Holmes, T. C. (2006). Electrical hyperexcitation of lateral ventral pacemaker neurons desynchronizes downstream circadian oscillators in the fly circadian circuit and induces multiple behavioral periods. *J. Neurosci.* **26**, 479–489.

Numano, R., Yamazaki, S., Umeda, N., Samura, T., Sujino, M., Takahashi, R., Ueda, M., Mori, A., Yamada, K., Sakaki, Y., et al. (2006). Constitutive expression of the period1 gene impairs behavioral and molecular circadian rhythms. *Proc. Natl. Acad. Sci. U.S.A.* **103**, 3716–3721.

O'Neill, J. S., and Reddy, A. B. (2011). Circadian clocks in human red blood cells. *Nature* **469**, 498–503.

O'Neill, J. S., van Ooijen, G., Dixon, L. E., Troein, C., Corellou, F., Bouget, F. Y., Reddy, A. B., and Millar, A. J. (2011). Circadian rhythms persist without transcription in a Eukaryote. *Nature* **469**, 554–558.

Ouyang, Y., Andersson, C. R., Kondo, T., Golden, S. S., and Johnson, C. H. (1998). Resonating circadian clocks enhance fitness in cyanobacteria. *Proc. Natl. Acad. Sci. U.S.A.* **95**, 8660–8664.

Peng, Y., Stoleru, D., Levine, J. D., Hall, J. C., and Rosbash, M. (2003). Drosophila free-running rhythms require intercellular communication. *PLoS Biol.* **1**, E13.

Peschel, N., and Helfrich-Forster, C. (2011). Setting the clock by nature: Circadian rhythm in the fruitfly Drosophila melanogaster. *FEBS Lett.* **585**, 1435–1442.

Peschel, N., Veleri, S., and Stanewsky, R. (2006). Veela defines a molecular link between crypto-chrome and timeless in the light-input pathway to Drosophila's circadian clock. *Proc. Natl. Acad. Sci. U.S.A.* **103**, 17313–17318.

Peschel, N., Chen, K. F., Szabo, G., and Stanewsky, R. (2009). Light-dependent interactions between the Drosophila circadian clock factors cryptochrome, jetlag, and timeless. *Curr. Biol.* **19**, 241–247.

Picot, M., Klarsfeld, A., Chelot, E., Malpel, S., and Rouyer, F. (2009). A role for blind DN2 clock neurons in temperature entrainment of the Drosophila larval brain. *J. Neurosci.* **29**, 8312–8320.

Price, J. L., Dembinska, M. E., Young, M. W., and Rosbash, M. (1995). Suppression of PERIOD protein abundance and circadian cycling by the Drosophila clock mutation timeless. *EMBO J.* **14**, 4044–4049.

Price, J. L., Blau, J., Rothenfluh, A., Abodeely, M., Kloss, B., and Young, M. W. (1998). Double-time is a novel Drosophila clock gene that regulates PERIOD protein accumulation. *Cell* **94**, 83–95.

Reddy, A. B., Karp, N. A., Maywood, E. S., Sage, E. A., Deery, M., O'Neill, J. S., Wong, G. K., Chesham, J., Odell, M., Lilley, K. S., *et al.* (2006). Circadian orchestration of the hepatic proteome. *Curr. Biol.* **16**, 1107–1115.

Richier, B., Michard-Vanhee, C., Lamouroux, A., Papin, C., and Rouyer, F. (2008). The clockwork orange Drosophila protein functions as both an activator and a repressor of clock gene expression. *J. Biol. Rhythms* **23**, 103–116.

Rosato, E., Trevisan, A., Sandrelli, F., Zordan, M., Kyriacou, C. P., and Costa, R. (1997). Conceptual translation of timeless reveals alternative initiating methionines in Drosophila. *Nucleic Acids Res.* **25**, 455–458.

Rothenfluh, A., Abodeely, M., and Young, M. W. (2000a). Short-period mutations of per affect a double-time-dependent step in the Drosophila circadian clock. *Curr. Biol.* **10**, 1399–1402.

Rothenfluh, A., Young, M. W., and Saez, L. (2000b). A TIMELESS-independent function for PERIOD proteins in the Drosophila clock. *Neuron* **26**, 505–514.

Rutila, J. E., Suri, V., Le, M., So, W. V., Rosbash, M., and Hall, J. C. (1998). CYCLE is a second bHLH-PAS clock protein essential for circadian rhythmicity and transcription of Drosophila period and timeless. *Cell* **93**, 805–814.

Saez, L., and Young, M. W. (1996). Regulation of nuclear entry of the Drosophila clock proteins period and timeless. *Neuron* **17**, 911–920.

Sandrelli, F., Tauber, E., Pegoraro, M., Mazzotta, G., Cisotto, P., Landskron, J., Stanewsky, R., Piccin, A., Rosato, E., Zordan, M., *et al.* (2007). A molecular basis for natural selection at the timeless locus in Drosophila melanogaster. *Science* **316**, 1898–1900.

Sathyanarayanan, S., Zheng, X., Xiao, R., and Sehgal, A. (2004). Posttranslational regulation of Drosophila PERIOD protein by protein phosphatase 2A. *Cell* **116**, 603–615.

Scheibler, E., and Wollnik, F. (2009). Interspecific contact affects phase response and activity in desert hamsters. *Physiol. Behav.* **98**, 288–295.

Sehadova, H., Glaser, F. T., Gentile, C., Simoni, A., Giesecke, A., Albert, J. T., and Stanewsky, R. (2009). Temperature entrainment of Drosophila's circadian clock involves the gene nocte and signaling from peripheral sensory tissues to the brain. *Neuron* **64**, 251–266.

Sehgal, A., Price, J. L., Man, B., and Young, M. W. (1994). Loss of circadian behavioral rhythms and per RNA oscillations in the Drosophila mutant timeless. *Science* **263**, 1603–1606.

Shafer, O. T., Levine, J. L., Truman, J. W., and Hall, J. C. (2004). Flies by night: Effects of changing day length on *Drosophila*'s circadian clock. *Curr. Biol.* **14**, 424–432.

Shafer, O. T., Helfrich-Forster, C., Renn, S. C., and Taghert, P. H. (2006). Reevaluation of Drosophila melanogaster's neuronal circadian pacemakers reveals new neuronal classes. *J. Comp. Neurol.* **498**, 180–193.

Shang, Y., Griffith, L. C., and Rosbash, M. (2008). Light-arousal and circadian photoreception circuits intersect at the large PDF cells of the Drosophila brain. *Proc. Natl. Acad. Sci. U.S.A.* **105**, 19587–19594.

Sheeba, V., Fogle, K. J., Kaneko, M., Rashid, S., Chou, Y. T., Sharma, V. K., and Holmes, T. C. (2008a). Large ventral lateral neurons modulate arousal and sleep in Drosophila. *Curr. Biol.* **18**, 1537–1545.

Sheeba, V., Gu, H., Sharma, V. K., O'Dowd, D. K., and Holmes, T. C. (2008b). Circadian- and light-dependent regulation of resting membrane potential and spontaneous action potential firing of Drosophila circadian pacemaker neurons. *J. Neurophysiol.* **99**, 976–988.

Sidote, D., Majercak, J., Parikh, V., and Edery, I. (1998). Differential effects of light and heat on the Drosophila circadian clock proteins PER and TIM. *Mol. Cell. Biol.* **18**, 2004–2013.

Siwicki, K. K., Eastman, C., Petersen, G., Rosbash, M., and Hall, J. C. (1988). Antibodies to the period gene product of Drosophila reveal diverse tissue distribution and rhythmic changes in the visual system. *Neuron* **1**, 141–150.

Skowyra, D., Craig, K. L., Tyers, M., Elledge, S. J., and Harper, J. W. (1997). F-box proteins are receptors that recruit phosphorylated substrates to the SCF ubiquitin-ligase complex. *Cell* **91**, 209–219.

Smith, E. M., Lin, J. M., Meissner, R. A., and Allada, R. (2008). Dominant-negative CK2alpha induces potent effects on circadian rhythmicity. *PLoS Genet.* **4**, e12.

Stanewsky, R., Frisch, B., Brandes, C., Hamblen-Coyle, M. J., Rosbash, M., and Hall, J. C. (1997). Temporal and spatial expression patterns of transgenes containing increasing amounts of the Drosophila clock gene period and a lacZ reporter: Mapping elements of the PER protein involved in circadian cycling. *J. Neurosci.* **17**, 676–696.

Stanewsky, R., Kaneko, M., Emery, P., Beretta, B., Wager-Smith, K., Kay, S. A., Rosbash, M., and Hall, J. C. (1998). The Cryb mutation identifies cryptochrome as a circadian photoreceptor in Drosophila. *Cell* **95**, 681–692.

Stoleru, D., Peng, Y., Agosto, J., and Rosbash, M. (2004). Coupled oscillators control morning and evening locomotor behaviour of Drosophila. *Nature* **431**, 862–868.

Stoleru, D., Peng, Y., Nawathean, P., and Rosbash, M. (2005). A resetting signal between Drosophila pacemakers synchronizes morning and evening activity. *Nature* **438**, 238–242.

Stoleru, D., Nawathean, P., Fernandez, M. P., Menet, J. S., Ceriani, M. F., and Rosbash, M. (2007). The Drosophila circadian network is a seasonal timer. *Cell* **129**, 207–219.

Storch, K. F., Lipan, O., Leykin, I., Viswanathan, N., Davis, F. C., Wong, W. H., and Weitz, C. J. (2002). Extensive and divergent circadian gene expression in liver and heart. *Nature* **417**, 78–83.

Suh, J., and Jackson, F. R. (2007). Drosophila ebony activity is required in Glia for the circadian regulation of locomotor activity. *Neuron* **55**, 435–447.

Sun, W. C., Jeong, E. H., Jeong, H. J., Ko, H. W., Edery, I., and Kim, E. Y. (2010). Two distinct modes of PERIOD recruitment onto dCLOCK reveal a novel role for TIMELESS in circadian transcription. *J. Neurosci.* **30**, 14458–14469.

Suri, V., Hall, J. C., and Rosbash, M. (2000). Two novel doubletime mutants alter circadian properties and eliminate the delay between RNA and protein in Drosophila. *J. Neurosci.* **20**, 7547–7555.

Sweeney, B. M., and Haxo, F. T. (1961). Persistence of a photosynthetic rhythm in enucleated Acetabularia. *Science* **134**, 1361–1363.

Taghert, P. H., Hewes, R. S., Park, J. H., O'Brien, M. A., Han, M., and Peck, M. E. (2001). Multiple amidated neuropeptides are required for normal circadian locomotor rhythms in Drosophila. *J. Neurosci.* **21**, 6673–6686.

Tang, C. H., Hinteregger, E., Shang, Y., and Rosbash, M. (2010). Light-mediated TIM degradation within Drosophila pacemaker neurons (s-LNvs) is neither necessary nor sufficient for delay zone phase shifts. *Neuron* **66**, 378–385.

Tauber, E., Zordan, M., Sandrelli, F., Pegoraro, M., Osterwalder, N., Breda, C., Daga, A., Selmin, A., Monger, K., Benna, C., *et al.* (2007). Natural selection favors a newly derived timeless allele in Drosophila melanogaster. *Science* **316**, 1895–1898.

Tomita, J., Nakajima, M., Kondo, T., and Iwasaki, H. (2005). No transcription-translation feedback in circadian rhythm of KaiC phosphorylation. *Science* **307**, 251–254.

Ueda, H. R., Matsumoto, A., Kawamura, M., Iino, M., Tanimura, T., and Hashimoto, S. (2002). Genome-wide transcriptional orchestration of circadian rhythms in Drosophila. *J. Biol. Chem.* **277**, 14048–14052.

Vosshall, L. B., and Young, M. W. (1995). Circadian rhythms in Drosophila can be driven by period expression in a restricted group of central brain cells. *Neuron* **15**, 345–360.

Vosshall, L. B., Price, J. L., Sehgal, A., Saez, L., and Young, M. W. (1994). Block in nuclear localization of period protein by a second clock mutation, timeless. *Science* **263**, 1606–1609.

Webb, A. B., Angelo, N., Huettner, J. E., and Herzog, E. D. (2009). Intrinsic, nondeterministic circadian rhythm generation in identified mammalian neurons. *Proc. Natl. Acad. Sci. U.S.A.* **106**, 16493–16498.

Welsh, D. K., Yoo, S. H., Liu, A. C., Takahashi, J. S., and Kay, S. A. (2004). Bioluminescence imaging of individual fibroblasts reveals persistent, independently phased circadian rhythms of clock gene expression. *Curr. Biol.* **14**, 2289–2295.

Wheeler, D. A., Hamblen-Coyle, M. J., Dushay, M. S., and Hall, J. C. (1993). Behavior in light-dark cycles of Drosophila mutants that are arrhythmic, blind or both. *J. Biol. Rhythms* **8**, 67–94.

Wijnen, H., and Young, M. W. (2008). The right period for a siesta. *Neuron* **60**, 943–946.

Wood, K. V. (1995). Marker proteins for gene expression. *Curr. Opin. Biotechnol.* **6**, 50–58.

Wuarin, J., Falvey, E., Lavery, D., Talbot, D., Schmidt, E., Ossipow, V., Fonjallaz, P., and Schibler, U. (1992). The role of the transcriptional activator protein DBP in circadian liver gene expression. *J. Cell Sci. Suppl.* **16**, 123–127.

Yamaguchi, S., Isejima, H., Matsuo, T., Okura, R., Yagita, K., Kobayashi, M., and Okamura, H. (2003). Synchronization of cellular clocks in the suprachiasmatic nucleus. *Science* **302**, 1408–1412.

Yamanaka, I., Koinuma, S., Shigeyoshi, Y., Uchiyama, Y., and Yagita, K. (2007). Presence of robust circadian clock oscillation under constitutive over-expression of mCry1 in rat-1 fibroblasts. *FEBS Lett.* **581**, 4098–4102.

Yang, Z., and Sehgal, A. (2001). Role of molecular oscillations in generating behavioral rhythms in Drosophila. *Neuron* **29**, 453–467.

Yang, Z., Emerson, M., Su, H. S., and Sehgal, A. (1998). Response of the timeless protein to light correlates with behavioral entrainment and suggests a nonvisual pathway for circadian photoreception. *Neuron* **21**, 215–223.

Yoshii, T., Sakamoto, M., and Tomioka, K. (2002). A temperature-dependent timing mechanism is involved in the circadian system that drives locomotor rhythms in the fruit fly Drosophila melanogaster. *Zoolog. Sci.* **19**, 841–850.

Yoshii, T., Heshiki, Y., Ibuki-Ishibashi, T., Matsumoto, A., Tanimura, T., and Tomioka, K. (2005). Temperature cycles drive drosophila circadian oscillation in constant light that otherwise induces behavioural arrhythmicity. *Eur. J. Neurosci.* **22**, 1176–1184.

Yoshii, T., Fujii, K., and Tomioka, K. (2007). Induction of Drosophila behavioral and molecular circadian rhythms by temperature steps in constant light. *J. Biol. Rhythms* **22**, 103–114.

Yoshii, T., Todo, T., Wulbeck, C., Stanewsky, R., and Helfrich-Förster, C. (2008). Cryptochrome is present in the compound eyes and a subset of Drosophila's clock neurons. *J. Comp. Neurol.* **508**, 952–966.

Young, M. W. (1998). The molecular control of circadian behavioral rhythms and their entrainment in Drosophila. *Annu. Rev. Biochem.* **67**, 135–152.

Yu, Q., Colot, H. V., Kyriacou, C. P., Hall, J. C., and Rosbash, M. (1987a). Behaviour modification by in vitro mutagenesis of a variable region within the period gene of Drosophila. *Nature* **326**, 765–769.

Yu, Q., Jacquier, A. C., Citri, Y., Hamblen, M., Hall, J. C., and Rosbash, M. (1987b). Molecular mapping of point mutations in the period gene that stop or speed up biological clocks in Drosophila melanogaster. *Proc. Natl. Acad. Sci. U.S.A.* **84**, 784–788.

Yu, W., Zheng, H., Houl, J. H., Dauwalder, B., and Hardin, P. E. (2006). PER-dependent rhythms in CLK phosphorylation and E-box binding regulate circadian transcription. *Genes Dev.* **20**, 723–733.

Yu, W., Zheng, H., Price, J. L., and Hardin, P. E. (2009). DOUBLETIME plays a noncatalytic role to mediate CLOCK phosphorylation and repress CLOCK-dependent transcription within the Drosophila circadian clock. *Mol. Cell. Biol.* **29**, 1452–1458.

Yu, W., Houl, J. H., and Hardin, P. E. (2011). NEMO kinase contributes to core period determination by slowing the pace of the Drosophila circadian oscillator. *Curr. Biol.* **21**, 756–761.

Zerr, D. M., Hall, J. C., Rosbash, M., and Siwicki, K. K. (1990). Circadian fluctuations of period protein immunoreactivity in the CNS and the visual system of Drosophila. *J. Neurosci.* **10**, 2749–2762.

Zheng, X., Koh, K., Sowcik, M., Smith, C. J., Chen, D., Wu, M. N., and Sehgal, A. (2009). An isoform-specific mutant reveals a role of PDP1 epsilon in the circadian oscillator. *J. Neurosci.* **29**, 10920–10927.

Zhu, H., Sauman, I., Yuan, Q., Casselman, A., Emery-Le, M., Emery, P., and Reppert, S. M. (2008). Cryptochromes define a novel circadian clock mechanism in monarch butterflies that may underlie sun compass navigation. *PLoS Biol.* **6**, e4.

Index

Note: Page numbers followed by "*f*" indicate figures, and "*t*" indicate tables.